EVERYDAY INNOVATION is a 'must read' for all those who wish to embed a culture of innovation in their businesses.
Prof. Brian MacCraith, President, DCU

Hugh is passionate about innovation and its role in delivering growth. In his book, he cuts to the core in explaining what involved in driving innovation and defining the simple practical steps you need to undertake as a business manager / owner. This book differentiates itself in that it creates an easy-to-understand guide and framework on how to effectively implement innovation in your business.
Sean McNulty, CEO, Dolmen / chairman, IMSC

Hugh captures the simplicity of the innovation management process yet succinctly details its importance in business today. A great read.
Donal Tierney, CEO, Bimeda / chairman, IRDG

At last, a book on innovation management that SMEs can refer to in order to put a structure on creativity so as to get repeatable success throughout their business.
Mark Fielding, CEO, ISME

EVERYDAY INNOVATION is a comprehensive yet easy-to-read book on setting up and operating an IMS in your business, irrespective of your size or sector. I particularly like the 'jargon buster' section, which helps to take the mystery out of innovation and explains the terms we have all heard (but were afraid to ask what they meant) in simple English!
Bernie McGahon, Inter*Trade*Ireland

... the book comes alive with practical tips gleaned from Hugh's own experience of leading innovation in a reluctant but co-operative organisation. Practical guides to such subjects as incentives, grants and tax reliefs, as well as pointers to recognised centres of best practice, make this a valuable companion for any business leader.
Gabriel D'Arcy, CEO, Town of Monaghan Co-Op

EVERYDAY INNOVATION

A Practical Guide to Establishing and Operating an Innovation Management System in your Business

Hubert Henry

·OAK·TREE·PRESS·

Published by OAK TREE PRESS
www.oaktreepress.com / www.SuccessStore.com

© 2015 Hubert Henry

A catalogue record of this book is available from the British Library.

ISBN 978 1 78119 180 4 (paperback)
ISBN 978 1 78119 181 1 (ePub)
ISBN 978 1 78119 182 8 (Kindle)
ISBN 978 1 78119 202 3 (PDF)

Cover design: Kieran O'Connor Design
Cover image: Stockbyte / thinkstockphotos.co.uk

The information contained in this publication is intended for guideline purposes only and does not represent legal advice. Readers should always seek independent legal and/or other professional advice specific to their own requirements before taking any action based on the information provided herein.

CONTENTS

FIGURES

TABLES

DEDICATION

To my Mum (Mary) and my late Father (Eamonn).

Also to Marie, Alex and Luke.

ACKNOWLEDGEMENTS

I am honoured to have Sir Richard Needham write the *Foreword* for the book. I first met Richard over 10 years ago when he was addressing a group of Enterprise Ireland companies on the need for an innovative approach in the clean tech sector. Wasn't he right! Throughout his career, Richard has always strived to bridge that well-known gap between invention and innovation. He has been one of the most successful leaders I know at this. Richard is also a former director of the very innovative Dyson group. Many thanks to him for his input.

Thank you to my employer, Bord na Móna plc, its chairman, John Horgan, its managing director, Mike Quinn, board of directors and senior management team. A special word of thanks to Gabriel D'Arcy, the group's former managing director, for writing the *Epilogue* to the book.

To my colleagues and friends of the innovation team and broader innovation network in Bord Na Móna. I hope we will progress the transformation journey we are on by continuing with our innovation agenda. A particularly big thank you to the following people who worked closely with me over the last six years and who helped fashion my thoughts on the subject: Helen Behan, Sean Creedon, Imelda Egan, Barry Hooper, Austin Lanham, Eamon Leigh, Sharon McGuinness, Ian Phillips, Pamela Ryan and Adele Woods. Your professionalism and commitment were the principal catalysts for me embarking on this project.

Thanks also to all my friends and colleagues in the wider Bord na Móna group, who also provided me with useful comments and insights on how the text could be improved. In particular, thanks to Michael Barry, Thomas Bradley, Pat Downes, Gerry O'Hagan, and Gerry Ryan. Thanks also to Nicola Joyce for supporting and putting up with me!

Thanks to all who reviewed the book drafts and provided much needed and useful comments on the text as it evolved. Also to all the busy people who provided me with insights on aspects of their very innovative businesses. In particular, thank you to Mark Bowkett (TellLab), Maurice Buckley (NSAI), Gary Carroll (EpiSensor), Seamus Connolly (Fastank Engineering), Mark Coyne (Dalkia / Veolia), Noel Crawford (IBM), Marie

Doyle-Henry (Orchid Consulting), Denis Hayes (IRDG), Linda Hendy (NSAI), Ciaran Herr (Horseware), Dr Tom Kelly (Enterprise Ireland), Dr Tony Lenihan (Fáilte Ireland), Rebekah Lyons (Portobello Institute), Simon Maddock (Diva Matrix Consulting), John McConnell (An Post), John McCurdy (Richard Keenan and Co Ltd.), Dr Bernie McGahon (InterTradeIreland), Sean McNulty (Dolmen), John McSweeney (ESB Group), Evelyn O'Toole (Complete Laboratory Solutions), Dr Anthony Owens (Arran Chemicals) and Catherine Wheelan (FBD Insurance).

A special mention of the dedicated staff in two state support organisations – Enterprise Ireland (EI) and the National Standards Authority of Ireland (NSAI) – is required. Both institutions, for many years, have promoted the development of an innovation culture in Irish business. Given Ireland's current standing in this area on the world stage, they have done a great job.

Many thanks to all the highly able staff from the third level institutions that participated in the Bord na Móna innovation ecosystem by carrying out R&D on our behalf and / or introducing external people or groups to our process.

To Oak Tree Press, the publishers – in particular, its CEO and the book's skilled editor, Brian O'Kane – for having the faith in me to publish this work and the patience to see it through to the end. I hope it will be worth it for you.

To a very special man, Nicholas Norris, for the extreme bravery and total selflessness he displayed. I am deeply indebted to him forever.

To my Mum (Mary), brothers (Ray and Ken), sisters (Lynda and Mary), my extended family and all my friends who have supported me throughout the project (and before!). I really appreciated your help even if I did not always show it. A huge thanks to all.

To my dearest wife, Marie, and our two terrific sons, Alex and Luke, who are both growing up be fine young men (despite my best efforts!), I offer my sincere thanks and express my deepest love for you all. You have supported, encouraged and frequently challenged me – especially when things did not always turn out as I had planned. For this, I will be eternally grateful.

Hugh Henry
January 2015

FOREWORD

When Hugh first approached me about writing a *Foreword* to his book, my initial thought was how to summarise all I wanted to say on innovation and business development in a few short pages? This was closely followed by a realisation that I first needed to introduce myself to the readers in order to set out my experiences in the area and hopefully demonstrate my passion for the subject. Well, here goes...

I am an ex-politician, who for much of my life also has taken a keen interest in business management and development. I spent over 16 years as director in Dyson Appliances Ltd and four years as deputy chairman there, where business development by diversification and innovation was a way of life. I like to think that I was influential in the rapid expansion and rise to prominence of the company during my time there. However, even before my time at Dyson, I was a keen business development advocate. Following a lengthy spell as a local representative on Somerset County Council, I entered mainstream politics in the UK House of Commons in 1979 (as an MP for Chippenham in Wiltshire). This occurred in conjunction with more than a few successful (and some not so successful!) stints in business. However, when I was promoted to Minister for the Economy in Northern Ireland (having previously served as private secretary to Jim Prior, the then Minister of State for Northern Ireland), I had to remove myself from my business interests and commit myself fully to public life. This I did readily, as devoting myself to public service was one of my earliest ambitions. During my very stressful but enjoyable time as Minister for the Economy for Northern Ireland, I worked hard to transform the economic base there to an export-led one under a national export promotion initiative. I was therefore in a good position to observe many businesses succeed and prosper while others floundered. What was the secret recipe to commercial success? This was something I promised myself I would explore once my political career came to an end. I was subsequently appointed UK Minister for Trade and Michael Heseltine's deputy, dealing with the entire UK's global export strategy, but my love for the commercial world was never far away.

Politics was a wonderful education for me. It was in fact a very effective and unforgiving *alma mater* in the way to get things done (what Hugh describes in the book as JDI!). After leaving politics, I was a much better and more astute business practitioner than when I entered it. Among the myriad of things I learned, two things stand out as being critical for my later successful career in business management. The first of these was how to be a good communicator (both written and oral). It is critical that one is able to articulate and / or document one's business proposition clearly and simply. The second is an ability to assimilate seemingly very disparate and often confused information and distil it into manageable chunks that then can be properly interpreted and communicated (identifying the 'wood among the trees').

I enjoyed my time in Northern Ireland immensely. It will always be in a place close to my heart (in fact, I have written a book on the subject, called *Battling for Peace*, an account of my years in Northern Ireland and my contribution to peace there). When eventually 1 left, I was Northern Ireland's longest-serving Minister but the time had come to move on and my passion for business development by commercial innovation was calling!

In the late 1990s, I became more and more involved in business. It was at this stage that I acted as a personal advisor and business development consultant to James Dyson. His company, a well-known technology company, was then examining the possibility of rapid growth by internationalising its activities. These were very exciting times as the Dyson group was expanding rapidly both from a market and NPD perspective but it urgently needed to be a more effective marketing, manufacturing, distribution and innovation management group. Just over a decade later, Dyson is a worldwide brand. It turns over in excess of £1.2 billion *per annum* and is a truly recognised and respected household name. How did a new market entrant, priced well above its competition and trading in a highly competitive sector (household appliances), do this? Well, it was not easy! It took a lot of hard work to get the group to where it is today. What I now see as innovation management, in the broadest sense of the phrase, was critical to achieving this. I am certain that innovation is essentially about managing one's business in a way that seeks and considers the contributions from a range of staff from different disciplines in a business and duly places all of these in a virtual 'matrix' for assimilation before deciding the best course of action. I am convinced that a good leader must be able to see these matrix relationships between well-meaning people of different specialties (NPD, manufacturing, sales / marketing, distribution, etc) and to act on them in the appropriate proportions. I like to refer to this as 'matrix innovation'. This was the key to success in Dyson and, on reflection, in many of the other successes I enjoyed in my career. One should not underestimate the gift of being a good mediator between different factions (yes, factions – in my time in business, my conflict resolution skills were often tested to the very limit!). The ability to listen first and act later is also one of the great traits of a good manager or innovator. People will get it wrong sometimes, but as I have said before: "Strong people make as many mistakes as weak people; the difference is that strong people admit their mistakes".

It is important to remember that less than 1 in 10 good products or services launched onto the market succeed. The effective management of the innovation matrix approach to the process is, to a large degree, the success or fail determinant. James Dyson managed to informally adopt this matrix approach very well.

Over the past decade, but in particular in the past few years, I have seen technology playing an increasingly important role in business innovation (as a technology maker, taker or user). I would be lost without my iPad or my smartphone, for example, despite the fact that I only really started to use either a few years ago. The rate of change is progressing so fast that I am convinced that, no matter what gadget we use nowadays, it will be totally obsolete in five years' time. Business leadership should take account of this, but management by matrix innovation will always be an important insulator to this change.

I am sorry to say that 'innovation' to me is becoming a much-abused term in business today. For this reason, I believe it is fast approaching its 'sell by date' as a term used in boardrooms throughout the world. As a veteran in the business sphere, I am now convinced that 'innovation management' and good old-fashioned 'leadership' are in fact one and the same. It is this matrix approach that is the special ingredient.

Therefore, in conclusion, I compliment Hugh on writing a simple to read and follow textbook on how we can all better manage our innovation process. I still believe that there will be a better chance of success for all involved if the main contents of this book are implemented with clear leadership from the very top and a matrix approach to management is adopted.

Best of luck to all readers. I wish you all great success in your endeavours.

Sir Richard Needham

1: INTRODUCING INNOVATION MANAGEMENT

Innovation distinguishes between a leader and a follower. (Steve Jobs)

Some initial thoughts

Innovation is the harnessing of ideas for the commercial benefit of a business. Such ideas range from small continuous improvements to larger cost-saving projects right through to the development of radical new approaches. Through these ideas, the primary aim of innovation is to drive future growth and profitability in a company. It is as simple and as grounded as that!

Business environments are changing so rapidly today that the useful life cycle of an enterprise is very short indeed. Less than 5% of the businesses that were in operation worldwide 75 years ago are trading today. Those that are have reinvented themselves, and have transformed their business or even moved into another sector! Nokia, the mobile telephone company, is a good example. Not many people know that Nokia started life in Norway as a wood pulp and rubber boots manufacturing company! Through a series of transformations, the company morphed into the world-class technology company that it is today. But even successful companies like Nokia are coming under pressure in the ever-changing smart phone / mobile phone market. Nokia recently was taken over by Microsoft and the future of the Nokia brand is now uncertain. Even the best companies must reinvent themselves again and again to survive.

Innovation is in some ways a pseudonym for change or transformation. It is only through constant evolution that our businesses will survive and prosper. Therefore, it is critical for any business to put innovation at the core of its activities, in order to survive in the long-term. However, this is easier to say than to do. In reality, we can all be so busy with the 'here and now' that we do not make time to consider new ideas or new ways of doing things. Consequently, innovation is often relegated to the 'rainy day' place, where we innovate only when there is nothing else to do. But innovation does not just happen in a business. It must be developed, nurtured and managed.

Innovation management in your business is a journey, not a destination. Never stop innovating! Businesses that stop innovating are likely to fail sooner or later. Remember Kodak? Never heard of them? Exactly! For readers under the age of 30, Kodak was once one of two giant corporations that dominated the global photographic film industry from the 1950s until the 1980s. Its principal product was photographic film that, once exposed, was developed in special camera shops (and later in pharmacies) into physical prints. It was only then that you could see which shots were good and which were blighted by red eye or a shaky hand. Believe it or not, this was the way we all took pictures as recently as 20 years ago! But while Kodak was busy concentrating its innovation efforts on outsmarting its main competitor (a Japanese company, Fuji), it failed to recognise how profound an effect digital photography would have on its business, despite having funded successful research into it! Digital photography has transformed the way we record pictures and video. Nowadays, most of us have very good quality picture-taking and storage capabilities on our smart phones and no longer need to use photographic film – or even cameras. Who saw that coming? To conclude the story positively, Kodak has regrouped and is emerging as a force to be watched in the professional printing area. How did it do this? By transformation through innovation, of course!

A formal innovation management system (IMS) simply brings a discipline to the process of creating ideas, picking the best of these and proceeding to develop the chosen ideas into marketable products or services, thereby improving your business. The main focus of innovation management should be to allow your organisation to respond to external or internal opportunities and subsequently to use its creative efforts to introduce new offerings to market. It is important that innovation is not left to the R&D department alone, which, as we will see, supports just one element of the process.

I passionately believe that formal innovation, in all its guises, is the best way for a business to remain agile and ahead of the competition in this challenging economic climate. It is the innovation of today that becomes the best practice of tomorrow. In view of this, every business should set out to leverage its particular commercial and technical strengths and, through applied collaboration or partnerships with specialist research institutions, capitalise on national and international market opportunities. The case for seizing this innovation opportunity is indeed compelling.

There will be detractors who believe that innovation is stifled by trying to manage it. I often hear people say:

- "That's already happening here, we just don't write the ideas down"
- "I do not have time to do this. I am very busy"
- "Why do you need a system in the first place?"
- "This is not innovation. It is business development or transformation".

But don't let comments like these deter you. Identify the key influencers in your business, get them on board and try also to get any detractors directly involved in the process. But don't spend too much time trying to get the detractors on side, at the risk of spending insufficient time with those who support your efforts. Successful innovation is the commercialised outcome of your efforts that will prove your critics wrong!

As we will see later in **Chapter 3**, only about 1% of ideas will ever be successful in the market place. If this is the first 1% of your efforts, then happy days! If not, the enthusiasm and commitment to continue can wane and the whole process can falter before it is given a realistic chance. Thus it is vitally important that innovation activities in a business are properly managed in order to yield the best results. This will only be achieved when you involve all staff at every level in contributing creatively to the company's future development.

Innovation management must be supported enthusiastically from the very top of the organisation and, unless there is buy-in from key influencers, it is certain to fail. At the start, do not overly complicate matters. In football or rugby parlance, get players on the pitch first; then make them better at what they do by giving them targeted training. Your team will be made up of players who do different things and therefore one size does not fit all when it comes to training. Take the time to identify what is relevant for your business first, but don't underestimate the importance of training. To succeed in building an innovation culture in your business, you need to be able to sustain innovation management for a long time. Remember, this is a marathon, not a sprint.

> Embrace the future and refer to the past but do not live exclusively in either. A wise man once said, "innovation without methodology is just luck"! Another said, "the harder I practice, the luckier I get"! Therefore, let's get a methodology in place and start practicing!

EVERYDAY INNOVATION is a no-nonsense book designed to help you take the plunge and invest in innovation and in the future of your business. It sets out 10 easy steps to establishing an innovation management system (IMS) in your business. It provides a comprehensive, but easy to follow, description of each step, and shows how to embed the IMS in your business. It explores the risks and potential pitfalls of installing an IMS and how to avoid them. In addition, the real-life comments throughout the book look at how companies close to home approach innovation. We also look at other countries that are considered to be exemplary in this area. The Glossary of Terms acts as a 'jargon-buster', explaining all the buzzwords and terms used in innovation management, in plain English! And the final chapter presents useful contacts to help you get started.

The book is written and structured in a way that it can be read by CEOs, sales or marketing practitioners, production managers, procurement professionals and technical people alike in order to make their approach to innovation management more complete. It is essentially an innovation management DIY toolkit!

Innovation defined – what it is, and what it is NOT!

As this book is all about innovation management, let's begin with a definition of innovation.

A simple web search for 'innovation' yields hundreds of thousands of references! 'Innovation definitions' yields over 100k references, while 'innovation management' yields over a quarter of a million!

Clearly, the term innovation means different things to different people. It is just as innovative to a company making widgets when a production line employee comes up with a better or faster way to produce the widget, as it is to put a rocket in space.

The best definition, in my opinion, is:

> **Innovation is the process of converting an idea or invention into a product or service that creates value which customers require and are prepared to pay for.**

No white coats, space rockets or flashing light bulbs there! Many people confuse creativity, invention and research and development (R&D) with innovation. I contend that there is a subtle but important difference between all these activities:

- **Creativity and innovation:** Creativity is thinking up new things; innovation is doing new things.
- **Invention and innovation:** Invention is the creation of a new concept; innovation is selling that concept.
- **R&D and innovation:** R&D is the conversion of money to knowledge; innovation is the conversion of that knowledge back into money.

Therefore, I believe that creativity, invention and R&D are all subsets or elements of innovation.

Finally, it is vitally important that, the outputs of innovation are followed up by bringing the results to the market and, at least, recovering the sunk costs associated with its development.

Why innovation management is important in any business

At the outset, it must be emphasised that you are innovating all of the time. However, in many cases, you just do not call it that or maybe you do not even know you are doing it! This is especially the case in SMEs, where the resources to commit to the innovation process simply are not there. I am convinced though that, by formalising the innovation process, you can make it better and that your business will benefit. This book is not advocating that you go out and hire or indeed commit scarce resources to the process. In my mind, it may well be counterproductive to do so. Likewise, it would be unwise if a business diverted resources from frontline customer services to the innovation process. However, if you do it right, you can start a formal innovation management process with little additional resource. As the saying goes 'start small but think big'. One does not need to be too focused on being too big too soon. Remember, getting the players on the field first is crucial. You can make the participants (or indeed yourself) better players later!

There are several reasons why the establishment and maintenance of an IMS can be critical to your business's success. A business that fails to innovate runs the risk of losing ground to competitors, losing key staff, operating inefficiently or even going out of business all together. Innovation can be a key differentiator between market leaders and their competitors. The old adage applies: 'fail to prepare, then prepare to fail'. On a grander scale, innovation also helps in the positive development of the country, society at large and our environment which makes it all possible. In short, we all benefit from innovation.

Innovation must form a key part of a business's overall vision and strategy. It is not only about designing a new product or service to sell and disrupt the market. It also can focus on existing business processes and practices to improve efficiency, find new customers, cut down on waste and increase profits (see **Chapter 5**). But above all else, innovation can give businesses a commercial advantage.

The benefits of innovation are many and include:

- Initial benefits (short-term):
  - Extend the product range in an existing market
  - Reduce fixed labour costs
  - Improve production processes
  - Reduce materials wastage

- ○ Reduce energy consumption
- ○ Improve quality (certification may be necessary)
- ○ Limit the environmental impact of activities (conform to regulations)
- ○ Replace dated products or services by upgrading the current product range
- ○ Create new markets for current offerings.
- Future benefits (medium-term):
  - ○ Keeps all employees focused on the business's vision and strategy
  - ○ Gives participating employees a role in where / how the business develops
  - ○ Retains key staff by creating a dynamic working environment
  - ○ Encourages business managers to empower employees to acquire the skills and ambition to succeed ('can do' attitude)
  - ○ Helps your customers, thereby strengthening your relationship with them (customer-led innovation)
  - ○ Allows business managers to be more agile in responding to emerging market opportunities
  - ○ Expands your business into new areas or new markets
  - ○ Achieves a greater market insight (listen to the voice of the customer and gain market insights)
  - ○ Prevents the business from becoming obsolescent
  - ○ Involves the outside world in the development of your business (an open innovation mode)
  - ○ Attracts more external financial support for your activities.

In addition, there are societal benefits (in the longer term) in developing an effective innovation management approach in an economy, where innovation:

- Assists in attracting new investment into the economy and transforming existing enterprises in the community to being bigger and better, thereby generating benefits for society at large
- Increases the availability of new commercial outputs for the venture capital sector to invest in, thereby creating a dynamic business environment in which all can participate
- Promotes the development of a national education system (at all levels) that encourages graduates to engage in lateral thinking. The resulting

creativity and inventiveness are areas that are vital in creating a smart, open economy

- Results in the state becoming actively involved in the acceleration of a dynamic business improvement culture through the provision of the required infrastructure and funding support

- Encourages the commercial sector in general to intensify its focus on the advantage of a strong innovative customer-centric approach with attendant benefit to the society at large

- Attracts more FDI investment to the country, while at the same time developing the indigenous community, thereby resulting in a more balanced economy.

One of the best innovation companies of our generation is Google. In an article in the Google *Think* newsletter (2011), Susan Wojcicki, then Senior Vice President of Advertising at the company (now CEO of YouTube), listed the eight key principles or pillars of good innovation management as:

- Have a mission that matters
- Think big but start small
- Strive for continual innovation, not instant perfection
- Look for ideas everywhere
- Share everything
- Spark with imagination, fuel with data
- Be a platform
- Never fail to fail.

Comment: The importance of innovation in our business

Company – Complete Laboratory Solutions, Galway

Employees – 40

Sector – Laboratory services in Ireland and the UK

Comment by – Evelyn O'Toole, CEO

Things were going along quite well in the early days when I first set up the company (1994), meeting my expectations at the time. When the company was six years trading, it was establishing a nice momentum with a client list to be proud of. We were making sales profitably but now I know that we were not paying sufficient heed to the old adage that 'cash is king'. With the benefit of hindsight, I can see that the business model we were adopting was very risky. We knew we were too dependent on one customer. We even conducted a risk assessment on our business to identify what impact it would have if they left us. Our analysis showed that we would survive, so we ploughed on investing even more in the business!

However, the worst happened and they left us a year later, after we had made a large infrastructural investment. Our overheads had increased, our general and administration costs had increased and these combined with the loss of an anchor client put us under real pressure. We haemorrhaged money for 10 months in a row.

I knew then that business model innovation and the operation of a balanced approach to our activities were urgently needed. To cut a very long and stressful story short, we started to think and behave as business owners for the first time, instead of as technical problem-solvers and, although we have come up against many challenges on the way, we have never looked back since. Nonetheless, the fear of the dreaded 'burning platform' returning has kept on our toes. Although we do not always recognise it as such, we now remain vigilant of imminent market developments and respond accordingly by appropriate innovations.

Our first port of call was to look at the market space in which CLS traded. Following the cattle BSE crisis, and the foot and mouth scare, which both had severe impacts on our food and environmental testing on which we were mainly focusing, we took the decision to take urgent action. The key component of our revision was to diversify into other target markets that would provide greater sustainability in the longer term. From this, we made a strategic decision to invest heavily into the medical device and pharmaceutical markets. We then looked at what products and services we could offer and whether we could introduce new initiatives and / or provide existing services more cost-effectively.

If I were asked to summarise the key benefits of us eagerly engaging in innovation or business model management, the following would be my measured response:

- It assists in the retention and motivation of key staff
- It helps retain and improve the quality of our service provided to customers
- It makes customer retention easier
- It ensures that we are focused on market changes and thereby promotes a culture of opportunity identification (products and markets)
- It forces us to listen to our customers
- It makes cost control in our current business easier
- It helps to balance the business risks (do not put all eggs in one basket)
- It provides for a more open and inclusive management style in our company
- It makes it easier to identify and obtain funding support for our efforts
- It makes us more willing and able to embrace change in our business.

In my experience, it is all about adapting to change and in some cases bringing it about (being a market leader). Innovation management gives us that edge and forces us to stay agile by being at least on par with our competitors and often well ahead of them. Customer retention is a critical consideration and listening to their changing needs is our passion (the voice of the customer).

It is not an exaggeration to say that the only reason we are in business today is that we were innovative in changing our business model in response to market needs. I am therefore a real fan of innovation management. However, the agility of being an SME means that we can respond to new innovations more quickly. As we grow, this may not necessarily be the case and we are aware of this too.

In summary, innovation is the difference between success and failure to us. We intend to continue to embrace it with vigour but in a very controlled and focused manner.

2: SETTING UP AN INNOVATION MANAGEMENT SYSTEM IN YOUR BUSINESS – QUICKLY

Never before in history has Innovation offered promise of so much to so many in so short a time. (Bill Gates)

Below is an outline of the 10 top tasks you need to do in order to get the process up and running. Simply implementing the key elements in establishing a formalised innovation management system (IMS) will go a long way to getting a working process in place.

The key principles of an innovation management system

In innovation management, one size does not fit all. However, there are a number of key principles that should be stated or be implicit in any system installed:

- **Operate a simple process:** The idea does not need to be simple, only the way of managing it
- **Be commercially focused:** Be aligned with the company vision and strategy
- **Be applied in all you do:** Be seen to be practical
- **Be led by firm market insights:** Be customer- or consumer-centric and market-led / -aware
- **Make sure the process is open to everyone:** Both inside and outside the business
- **Recognise and reward participants:** Recognition is critical; the best reward is to share any intellectual property (IP) arising from the initiative

- **Everything you do must achieve the sustainability criteria laid down in the business's strategy:** The triple bottom line is a concept gathering traction in business
- **Celebrate success but allow for failure:** Only about 1% of what you do will make it to market
- **Communicate successes enthusiastically:** Internally and externally
- **Innovation is a team process:** It happens best when people with different skill sets and perspectives interact together
- **Make sure that all you do is relevant:** If innovation is not relevant to what you are doing, you may simply be creating a Chindōgu-type invention.

Figure 1: Examples of Chindōgu innovation in Japan

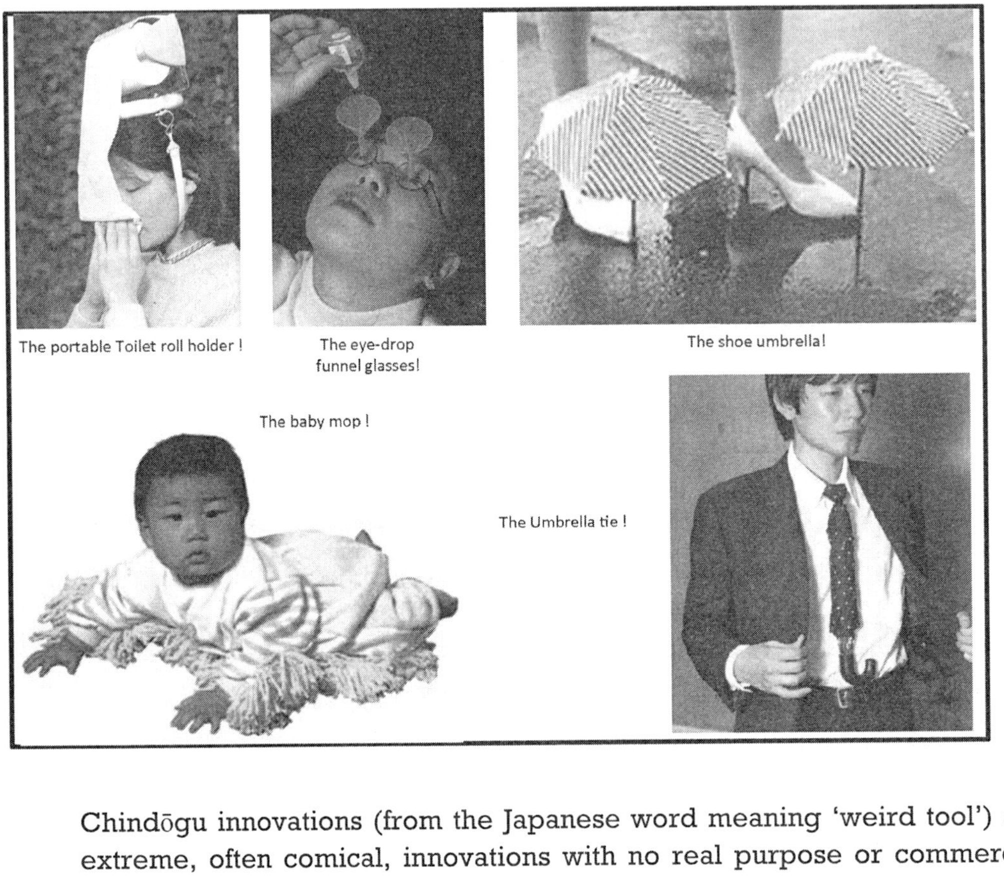

The portable Toilet roll holder !

The eye-drop funnel glasses!

The shoe umbrella!

The baby mop !

The Umbrella tie !

Chindōgu innovations (from the Japanese word meaning 'weird tool') are extreme, often comical, innovations with no real purpose or commercial application. But occasionally, the passage of time provides a purpose – for example, the 'selfie stick' was once a chindōgu innovation!

10 steps to establish an innovation management system

This section presents the 10 main steps that any business needs to follow in establishing an IMS. It is not, however, a substitute for reading and implementing the detail provided in the following chapters. It is merely meant to be a 'fast-track' to getting a system in place. The approach summarised here will get you started; the later chapters 'fill in the gaps' by providing more detail.

Step 1: Get senior management support for the process

Main things to do:

- The tone comes from the top, so get the support you need from the CEO or business owner

- Be clear what the business means by innovation and communicate this to all staff

- The IMS must be clearly aligned to the general business vision and strategy of the business

- The case for your business adopting an innovation management approach must be seen as compelling by the senior management team; it is your job to convince them of this

- Get the key influencers in your business to buy into the process; identify the main detractors and try to change their mind – but do not spend too much time on this.

Step 2: Establish the team and the innovation ecosystem

Main things to do:

- Pick the right people for the job: drive, discipline, influence, respect for others and enthusiasm are key qualities – a multidisciplinary team is needed

- All innovation must be based on market insights, so choose your team with this in mind

- Train the team in the appropriate tools to use to make them better innovators – but get them playing first

- Innovation management oversight should be the responsibility of one senior manager who reports directly to the business leader; and innovation must be a key element of that person's annual performance appraisal.

Step 3: Establish the correct structure for the team

Main things to do:

- Decide what suits your business best: an open or a linear / closed system
- Choose what structure fits best in your business from the centralised, de-centralised or hybrid models
- Establish an innovation framework and innovation matrix to suit your business
- Develop an IP management strategy, if applicable (mainly relates to product innovation)
- Introduce a rewards and recognition system in your business.

Step 4: Set objectives to suit your business

Main things to do:

- Communicate your intentions widely across the business but do not overpromise – you need to manage expectations of all involved, remembering that only 1% of what you do will be commercially successful
- Establish what you want to get out of the process and what is required to do it
- Use tools like PESTEL and SWOT to assist you in this process
- Set SMART objectives for the team
- Establish the key critical success factors (CSFs) in meeting these objectives.

Step 5: Establish what resources are to be committed to the process

Main things to do:

- Typically this is a % of turnover / revenues appropriate for your sector, ranging from 1% to 15%
- Your success or otherwise in the process invariably will be tied into the cost / benefit of the investment made (typically measured by return on investment (ROI)) – so concentrate on early wins.

Step 6: Design the system

Main things to do:

- Put a system in place for identifying opportunities (ideas management)
- Put a system in place for picking the best opportunities (ideas prioritisation)
- Run with the best projects identified but put a system in place to review project progress along the lines of 'failing quickly and failing cheaply' (review gate and project management disciplines)
- Get the innovation to the market to start it generating returns ASAP so that a return on investment (ROI) can be demonstrated (exploitation)
- Make sure there is a good balance between your short-term innovation activities and the 'further out there' projects (balanced portfolio)
- Accredit the system to an internationally recognised standard, if possible
- Lots of things can go wrong for you, so manage the risks by using a risk management process to help keep them under control.

Step 7: Measure, quantify and communicate the outputs

Main things to do:

- Establish the value of the process by visibly reporting on a series of well-chosen performance indicators (KPIs)
- Show the value of your innovation process by tracking the value of your innovations
- Develop a tracking matrix, such as the Innovation Value (IV) approach
- Communicate successes to the business but tolerate failure.

Step 8: Establish and maintain an innovation culture

Main things to do:

- The adherence to a process for innovation in a business leads to an innovation-centric behaviour in staff and, eventually, when it becomes embedded in a business (is part of its DNA), an innovation culture will develop. Do not be too impatient, it will happen!
- Involve as many people in your innovation ecosystem as possible; include stakeholders outside your immediate business (open innovation)
- Recognise and reward participants appropriately

- Hold annual innovation seminars or fairs for the participants in your ecosystem in order to showcase successes.

Step 9: *Obtain as much external financial support as you can*

Main things to do:

- Obtain maximum direct grant funding
- Obtain maximum tax credits or refunds
- Develop a resource tracking system to facilitate funding claims.

Step 10: *Review the entire process regularly*

Main things to do:

- Report regularly using selected innovation key performance indicators (KPIs)
- Monitor the performance of the system against the agreed critical success factors (CSFs)
- The initial focus must be to achieve 'early wins' in order to demonstrate value of the initiative
- Manage expectations so as to avoid loss of confidence in the process
- A performance dashboard can help you track how you are doing
- It is vital that you regularly refresh the process! The process must remain dynamic to employees and external partners always.

> **Remember, getting the players on the field first is crucial. You can make the participants (or indeed yourself) better players later!**

Comment: Generating and maintaining an innovation management culture in a large established company

Company – ESB Group, Dublin

Employees – >10,000

Sector – Energy generation, transmission and management

Comment by – John McSweeney, Director of Innovation

Most people do not think about where electricity comes from, how it is generated or the way in which it is brought into our homes. Although we take it for granted, having a secure, affordable and sustainable electricity supply is becoming increasingly difficult to sustain.

The scale of these challenges and opportunities facing the energy sector requires new thinking and innovative solutions. New technologies, greater competition and an increasingly sophisticated consumer mean that ESB must innovate faster to remain competitive. In response to this, a dedicated innovation business unit was set up to be the focal point to exploit new ideas that drive growth opportunities and offer solutions to the complex challenges that the energy industry faces. The innovation strategy developed by my team charts a roadmap and a structured approach that is focused on three key areas:

- Exploring new and emerging energy technologies
- Embracing the innovative strength of our people
- Collaborating more effectively across the company, with our customers, businesses and R&D partners.

Exploring new and emerging energy technologies

Almost every other industry has been affected by some form of disruptive technology and, in some cases, entire sectors have been wiped out. Nothing like that has happened in the electricity sector but the rate of change is happening so fast that there will undoubtedly be a disruptive force in electricity in the future. Solar power is perhaps next on the horizon. Our role is to be aware of new developments like this and to play our part in developing them to ensure they are part of the company's future plans. The Emerging Energy Technology group is playing a leading and innovative role in the development of solutions for the energy system of the future such as ocean energy, solar photovoltaic, telecoms and fibre optic broadband and electric vehicles.

ESB recognises that the new solutions in technology will come from a variety of sources. To that end, the €200m ESB Novusmodus fund was set up to invest in a wide range of new businesses in emerging and often breakthrough technologies.

Looking forward and adapting is our way of life and our priorities are to continue the innovation that has been at the centre of the company for generations.

Collaborating more effectively across the company, with our customers, businesses and partners

ESB views collaboration and partnerships with enterprises, representative groups, universities and other utilities as an important contributor to the development of future technologies, products and services.

Another pillar of our Innovation strategy is to develop even stronger relationships with our partners to create new opportunities and to commercialise new initiatives. Most notably this year, we have signed a joint venture agreement with Vodafone to roll out fibre broadband on ESB's network to more than 500,000 homes and businesses, which will revolutionise the Irish broadband experience. Other successful collaborations include an innovative energy agreement with Dublin Airport Authority, targeting a 30% energy saving at Dublin Airport, and a smart electric car charging IT system with IBM.

3: WHAT TO FOCUS YOUR EFFORTS ON

*The key to success is for you to make a habit throughout your life of
doing the things you fear. (Vincent Van Gogh)*

There are a number of basic steps to install and maintain an innovation management system (IMS) in your business. There is no magic formula to follow. However, in summary, an IMS is about collecting ideas, prioritising them and then setting up a review gate, project tracking and project management system.

Innovation is not all about technology or flashing light bulbs! If I had a euro for every time someone asked me, "What is the next big thing you are coming up with?", as I said what I did for a living, I would indeed be a wealthy man and I would not have to work for a living any more. Now that would be true innovation!

Unfortunately, the term 'innovation management' has become synonymous with the 'knowledge economy', which has widely been mooted as a way to get this great world economy of ours out of the recession we are currently in. However, I am frequently amused (and I admit, sometimes frustrated) by the widely-held misconception that the knowledge economy will be achieved by someone else making a life-altering discovery, a new world wide web-type of invention, a new wheel or an iPod / iPad replacement perhaps. Many people still seem to believe that someone somewhere out there is close to finding a 'pot of gold' idea that will revolutionise the world markets and possibly then the good times will come roaring back. Wishful thinking!

The game-changing discovery may indeed happen but it is through the piece-by-piece or incremental innovation (called 'Kaizen' innovation – from the Japanese word meaning 'improvement'), where small process improvements and business model improvements are made, wastage avoided and continuous new product / service development activities practiced, that the best results will be seen.

The idea that innovation is the domain of white-coated scientists or engineers looking down microscopes waiting for a 'eureka' moment could not be further from the truth. It is not the remit of 'others' to come up with all the solutions. We are all innovators and all have a key role to play in the process!

There is much more to innovation than technical innovation. On the contrary, multiple types of innovation can be identified, including:

- Strategic innovation
- Business model innovation
- Customer-led innovation
- Technology-led innovation.

However, it is widely felt that the end has come to product innovation and that innovation now is more about business model innovation based on consumer experience. Thus, technology is merely seen as an enabler in this process.

Selecting an appropriate innovation framework and matrix

In innovation management, no one size fits all and it is an individual's or group's choice which framework best suits their business. Below are a number of different models that I have seen implemented. Selection of the most appropriate for your business depends on many factors including:

- The people you have available
- The model chosen
- Your business's % market share
- The rate of change in your market / sector
- The competitors your business faces.

An innovation framework

An innovation framework is a description of the key elements of an IMS – specifically, the capture and prioritisation of ideas followed by the delivery of projects and, ultimately, the market exploitation of the resulting outputs. A framework recognises that many inter-dependencies exist between different disciplines involved in the process and therefore different resources are needed to get results.

There are four critical elements to any successful IMS that are represented in a framework. Take special notice of these in the review process (**Chapter 11**):

- Senior management and the 'boss' must be fully on board with the initiative and must stick with the process as it evolves
- A business must ensure that its vision and strategy is fully aligned to the innovation system
- You need the right people to deliver on the innovation business objectives
- Is the process you have in place working and is it being adhered to?

The innovation framework can be depicted in several ways. **Figures 2** and **3** show two of these ways.

Figure 2: Framework showing the key elements required from stakeholders in establishing an IMS

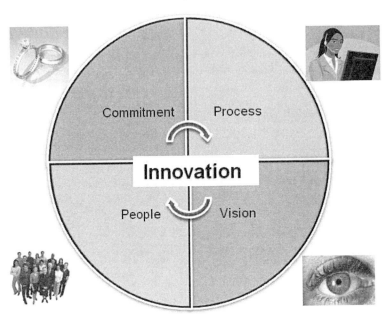

Figure 3: A concentric circle framework of the key components of an IMS

An innovation matrix

An innovation matrix is a tabular or schematic representation of where to focus your efforts to yield the best results. The matrix is simply the what and where of innovation – **what** to look for and **where** to look for it.

The matrix below shows the 'What' options you face. It is a description of where you are with your offering in the market place and therefore summarises what you need to focus on in order to bring your product / service to where you want it to be. Four options are identified as follows:

- **Incremental innovation:** Usually an existing offering in an existing market
- **Disruptive innovation:** Usually a new offering in an existing market
- **Semi-radical innovation:** Usually a new offering in an existing market
- **Radical innovation:** Usually a new offering in a new market.

Figure 4: A quadrant-based matrix: What to focus on when installing an IMS

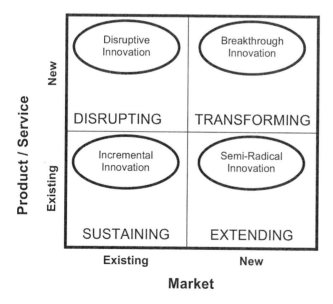

Establish what you want out of the process

Do not worry too much about nailing down precise, all-encompassing systems at this stage. The key is to ensure that the structure of your innovation system and its design are fully aligned with the strategy of your company (not the other way around, as is often the case).

It is often very useful to sit back and look at your business, where it is going and what macro and micro influences will impact on it in the coming years. It pays to spend a while looking at your business's vision and how you might achieve it before you rush into putting structures in place (including establishing an IMS). There is a definite hierarchy you should follow:

Vision ➡ **Strategy** ➡ **Structure**

Many tools can be used in describing the business you are in now and predicting how certain factors will influence it in the coming years. Two of these are:

- **PESTEL:** This is a useful tool to identify the current and future Political, Economic, Social, Technological, Environmental and Legal issues (positive and negative) likely to impact your business. The PESTEL overview of your business provides much of the information for the next stage in the process: the SWOT analysis

- **SWOT:** This is a simple documentation of the Strengths, Weaknesses, Opportunities and Threats of / to your business. Together with the outputs of the PESTEL analysis, the SWOT should be used to develop your business's innovation management objectives.

Figure 5: PESTEL and SWOT templates for your IMS

POLITICAL	ECONOMIC	SOCIAL
Government stability / initiatives Taxation policies / incentives Privatisation policies Regulation policies Social welfare policies	Business / economic cycles GNP trends Interest rates Unemployment Inflation Disposable income	Attitudes to work Lifestyles Demographics Education Social mobility
TECHNOLOGICAL	ENVIRONMENTAL	LEGAL
New technologies Improvements / developments Information technology Basic research investment Technology transfers	Waste disposal Pollution control Transport Energy supply Spatial planning Climate change Sustainability	Health & safety Employment law Monopolies legislation Trade restrictions

Strengths	Weaknesses
Opportunities	**Threats**

Decide how much to spend on the process

Table 1 summarises how much different industries spend on innovation management each year. The variations are great but reflect the nature of the markets in which they trade.

Table 1: How much some business sectors spend on innovation

Sector	% turnover spent on innovation
Technology	5-10
Pharmaceutical	15-20
ICT	10-15
Food	3-4
Utilities	1-2
Diversified	2-4
SMEs	1-10

The idea that innovation is the domain of white-coated scientists or engineers looking down microscopes waiting for a 'eureka' moment couldn't be further from the truth. It is not the remit of 'others' come up with all the solutions. We are all innovators and all have a key role to play in the process!

Design a system to suit your business

When you have established the appropriate innovation structure for your business and ensured that it is fully compatible with the goals and strategy of your business, it is good to consider the basic innovation system you wish to install.

Key components of the system

Essentially, there are four main elements:

- Identify the opportunities (ideas or opportunity management)
  - Ideas generation
  - Ideas prioritisation
  - Ideas investigation (the 'quick look' followed by the business case)
- Project delivery (review gate and project management disciplines)
- Getting the return (exploitation)
- Getting the balance right (balanced portfolio).

These components are shown diagrammatically in **Figure 6**.

Figure 6: The innovation cycle

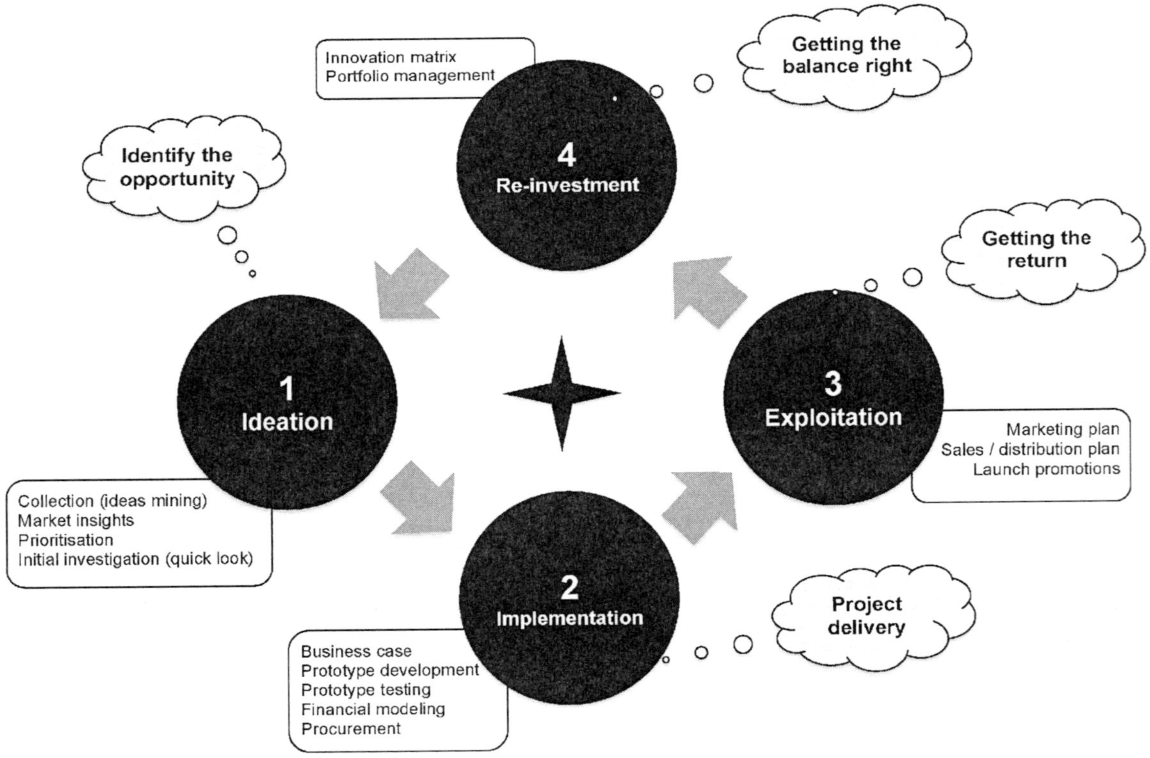

Identify the opportunity

1 Ideation

Collection (ideas mining)
Market insights
Prioritisation
Initial investigation (quick look)

2 Implementation

Business case
Prototype development
Prototype testing
Financial modeling
Procurement

Project delivery

3 Exploitation

Marketing plan
Sales / distribution plan
Launch promotions

Getting the return

Getting the balance right

4 Re-investment

Innovation matrix
Portfolio management

The triple bottom line concept

The 'triple bottom line' phrase was coined in 1997 by John Elkington, and is used liberally in corporate reports nowadays, often alongside the word 'sustainability'. This is because there is now an acceptance that innovation and sustainability go hand in hand. The message is:

> We are running our businesses this way because we wish to sustain our enterprise (Profit). We cherish the staff members who make this possible (People). And finally, we conduct our affairs so that we are not harming the very environment that makes this possible (Planet).

Figure 7: The triple bottom line

The 'triple bottom line', ideally, transforms a business into one that is less impacted by the external factors such as a global financial crisis or the weather (climate change) and operates under the general banner of sustainability. This is indeed a 'holy grail'.

Comment: What should we focus on and why?

Company – Portobello Institute, Dublin

Employees – 20 FTEs with an additional 100 tutors

Sector – SME in the education sector

Comment by – Rebekah Lyons, CEO

We are in the business of offering education, mainly early stage education (L4 to L6 on the HEA scale). We only operate in a small niche area of the education market but in our area we are close to market leader.

We trade in a very competitive market sector and the primary reason for our success to date has been that we have listened to the voice of our customer and avoid competitive 'turf wars' by not going head to head with more recognised educational awarding brands. We respond to market changes quickly, which is a useful feature of our SME status. Therefore, our focus is primarily on the 'here and now' or immediate term innovation because if we do not act here, we will not be in business in a very short time indeed. We are constantly reinventing our unique selling proposition (USP) and to do this we try to put ourselves in the shoes of the student.

Initially, I would have said we do not innovate but, on reflection, there is no doubt in my mind now that we do. We are a totally different business today than we were only five years ago. Our transformation was very much in response to changes in the market place. As we were not in the business of making 'widgets', we frequently were guilty of overlooking the fact that we were innovating, preferring to call it strategic marketing or business development.

The things we focused on in our business development have been market-led and responsive rather than proactive or future-looking. The driver has always been a burning platform of going out of business due to overdependence on one or a small number of customers or cost structures being wrong in the business. Our management style is informal and currently resides mainly in the head of the business leader. Recent changes in the market resulted in a significant chunk of the business's revenue effectively being lost overnight. The business responded by generating a new market for its current offering and later a new product for this market. The movement of the business into new market areas has accelerated this customer de-risking process. Therefore, an informal balanced portfolio management process is being followed.

I clearly recognise that the tone from the top is a critical determinant to our on-going strategy development. I personally am passionate about the development of our small enterprise. This poses an issue for succession planning since, if I was removed from the picture, the business may be less willing or able to adapt to market changes. This issue is top of our agenda to address, as our very survival depends on it.

While the main element of our innovation management (or strategic marketing) focus in the immediate term will be mainly in the market response (short-term) or follower mode, we do realise that intellectual property or intangible assets, including patents, trademarks and copyright, will play an important role in the future of the business. In addition, we recognise that the future-proofing of the business is important. However, for the foreseeable future, the innovation activities in our business, by necessity, will be in the 'here and now' sphere.

4: WHO SHOULD INNOVATE?

The reasonable man adapts himself to the world; the unreasonable
one persists in trying to adapt the world to them. Therefore, all
progress depends on the unreasonable man. (George Bernard Shaw)

The innovation groups in your 'ecosystem'

Innovation in a business cannot – and should not – be delegated. The drive
to innovate begins at the top. If the CEO does not protect and reward the
process, it will fail. So many CEOs either 'do not get innovation' or cannot
get past their finance-focused mindsets. If you cannot get the CEO
engaged around innovation, the odds against innovative success are
extremely high. Great innovation inevitably requires great internal
'selling' of the idea in order to build political support among top
management.

It is vital that you get senior management support for the innovation
process. The leader of your business must really believe and demonstrate
that an innovative way of doing things is a core consideration in his / her
business strategy and therefore forms an integral part of 'the way things
are done around here'. They must be acutely aware that the innovation of
today becomes the best practice of tomorrow. From business model and
customer-led innovation to technology development / utilisation initiatives
(most businesses are mainly technology-takers not technology-makers!),
everyone must use the appropriate tools to deliver on the business's
commercial objectives. Where possible, leverage the commercial and
technical strengths within the organisation augmented with targeted
collaboration / partnerships with external specialists to capitalise on
national and, where appropriate, international market opportunities.

The case for adopting an innovative approach is compelling and therefore
is at the core of what your business must do. Middle management is the
key ally of any innovation process. Managers are not natural champions of
innovation. They are often judged on short-term results, like monthly

financial performance, and therefore are slow to adopt new ideas in favour of short-term efficiency. I have often heard managers say that "it is not the number of balls you juggle in the air that you are judged on, it is the one you let fall!". This statement summarises well the conservative nature of middle managers in business.

It is not that middle managers are obstructionists or lack vision. It is just that the entire culture of most successful organisations is focused on the here and now, and rewards for managers are effectively (albeit unintentionally) designed to resist innovation. In the battle, it is unfortunate that short-term efficiency wins over innovation almost every time.

Table 2: The organisation of an innovation management ecosystem

Group	Members	Role	Meeting
Tier 1	Core Innovation Team	The core innovation management team's job is to assist, record, collaborate and oversee the process.	Weekly
Tier 2	Core Innovation team + Innovation family	This team's key role is to select people in their department to carry out required activities and to identify, liaise and deliver the innovation agenda.	Monthly
Tier 3	Core Innovation team + Innovation family + Innovation community	These are the important commercial and marketing people in your business, as well as the technical people not involved on a day-to-day basis. It is referred to as the innovation community.	Quarterly
Tier 4	Core Innovation team + Innovation family + Innovation community + Innovation network	This is the entire group of participants involved in your innovation process (both internal and external). The group's main function is simply to participate in the process.	Annually

Figure 8: The innovation management ecosystem

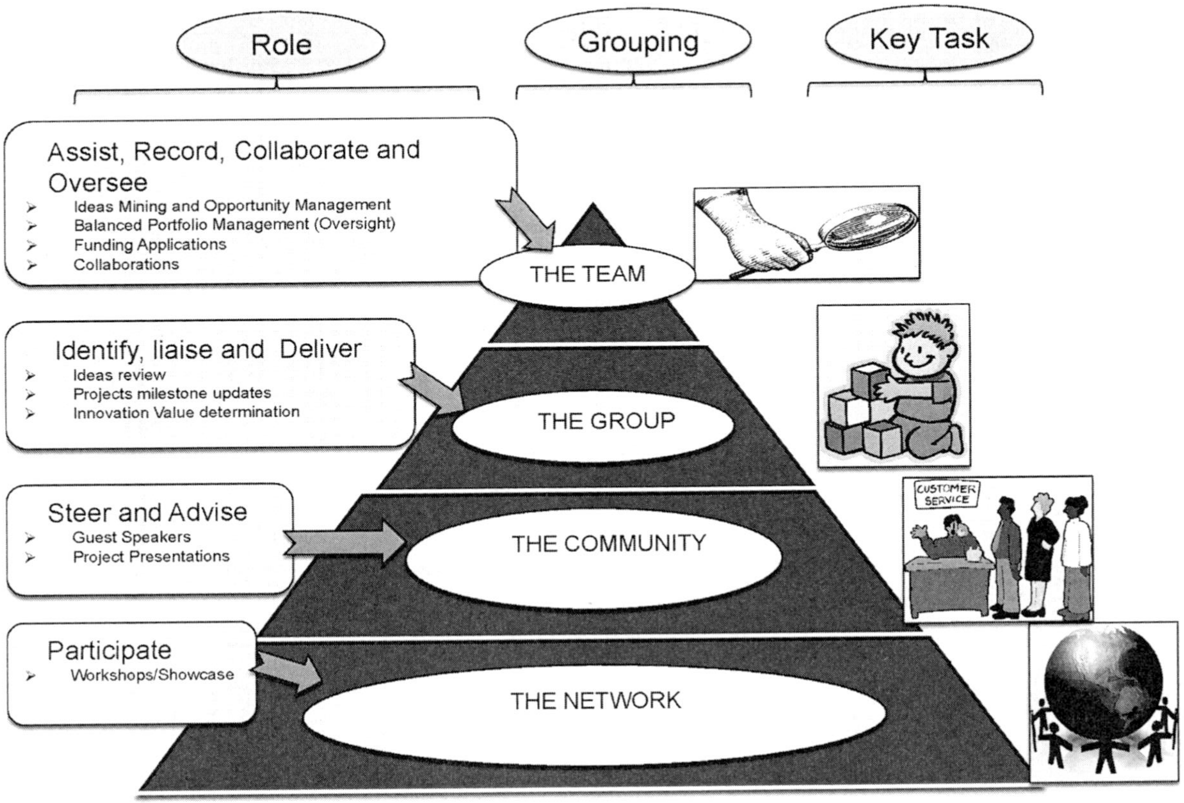
Role
Grouping
Key Task
Assist, Record, Collaborate and Oversee
Ideas Mining and Opportunity Management
Balanced Portfolio Management (Oversight)
Funding Applications
Collaborations
THE TEAM
Identify, liaise and Deliver
Ideas review
Projects milestone updates
Innovation Value determination
THE GROUP
Steer and Advise
Guest Speakers
Project Presentations
THE COMMUNITY
CUSTOMER SERVICE
Participate
Workshops/Showcase
THE NETWORK

Innovation structures

Non-structured innovation management (the informal model)

This structure – operated by most Irish businesses (particularly SMEs) – more accurately is an un-managed (free for all) system. It generally involves an extreme lack of process control with attendant inefficiencies and an overall lack of alignment to the business strategy (an 'it will be alright on the night' attitude). In young companies, this system often works well but it needs to progress to a more developed and managed structure as the company grows.

Figure 9: Non-structured (informal) innovation management

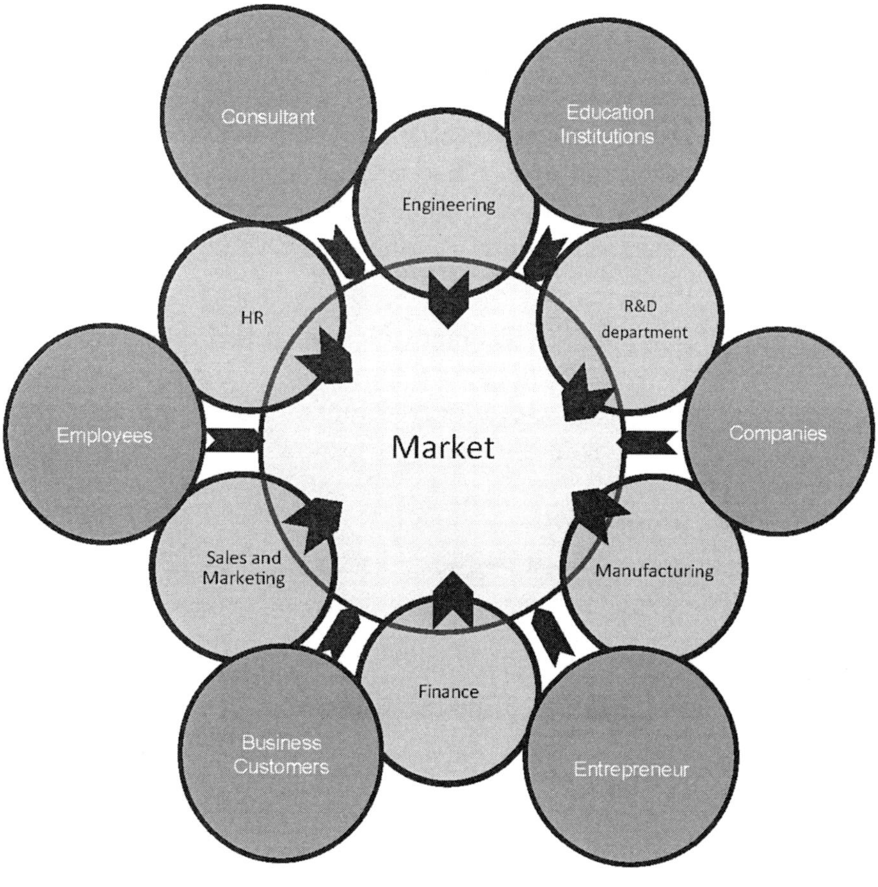

There are many disadvantages to operating this system, not least being that there are often multiple contact points with the market, leading to confusion and often duplication of effort (the left hand does not know what the right hand is doing!). In addition, under this approach, innovation becomes a 'rainy day' activity and when things are busy it is sent to the back of the queue not to be progressed again until things are a bit less hectic!

Structured innovation management

It is now widely accepted that this is the way to organise a business if you want to get the most from your innovation efforts. In structured innovation, two main drivers are identified:

- Technology push
- Market pull.

A push process is based on existing or newly-invented technology that the organisation has access to and tries to find profitable applications to use. A pull process tries to find areas where a market's or customer's needs are not met and then focuses development efforts to find solutions to those needs.

In these two main types of structured innovation, a closed system involves only the people within your organisation and follows a somewhat linear pattern as opposed to a wider or open system that involves others outside your organisation in a more rounded web-like structure.

Closed and linear innovation

This process is often linear in that there is a recognised beginning, middle and end to the work. Three main steps are typically recognised:

- Invention (dream it up)
- Development (make it happen)
- Diffusion (sell it).

Figure 10: Linear innovation

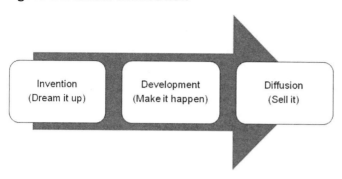

Closed and linear innovation is the way it used to be and, for many businesses, this remains the model of choice – for example, in much of the pharmaceutical, ICT and many other sectors where, understandably, innovations or improvements often are closely guarded from the competition. However, this can limit the degree to which these businesses engage with the outside world. This is the single most referenced drawback of this model.

Open innovation or collaboration

This is the modern way of innovating. It expands the reach of your IMS outside the boundaries of your own business's employees. Open innovation assumes that firms can – and should – actively seek out and use external ideas as well as internal ideas, and internal and external paths to market, as they look to advance their technology. Alternatively, it can be defined as 'innovating with partners by the sharing of risk and the sharing of reward'.

The central idea behind open innovation is that we live in a world of widely-distributed knowledge, in which companies cannot afford to rely entirely on their own research, but should instead buy or license processes or inventions (patents) from other companies. In addition, internal inventions not being used in a firm's business should be taken outside the company (through licensing, joint ventures or spin-offs) for exploitation. The open innovation approach should go beyond just using external sources of innovation such as customers, rival companies, and academic institutions to include more radical change in management by sharing the intellectual property outputs of innovation practices with the outside world.

Figure 11: The open innovation model

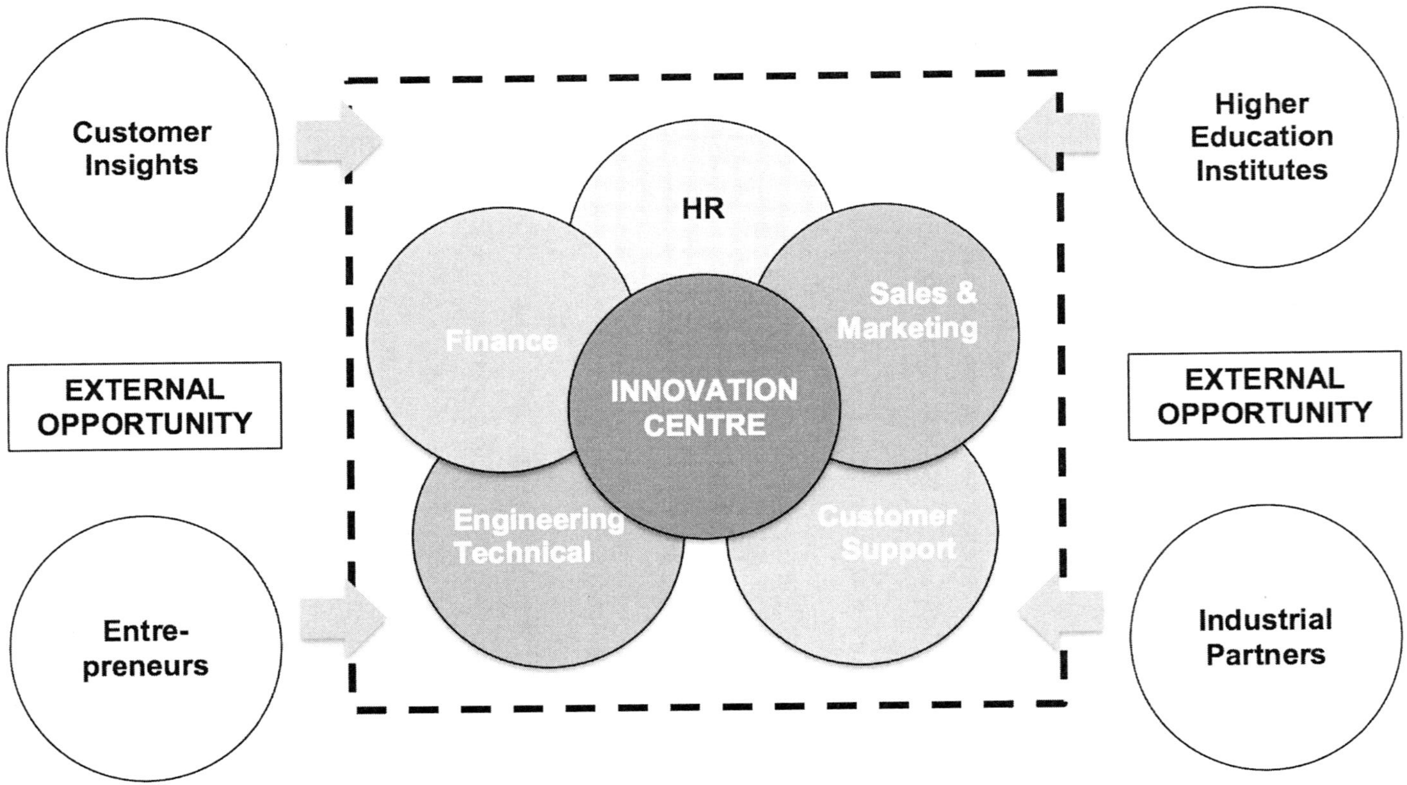

Choosing the appropriate structure for your business

There is no simple answer to the appropriate choice of structure for a business. It depends on many factors, such as:

- Maturity of your existing market and its rate of change
- Competitors
- Your presence in the market place (dominant or small player)
- Resources available (people and financial)
- Company strategy and ambition.

If you decide to establish a structured innovation management process in your business (and I strongly recommend you do) as opposed to a fully outsourced or an unmanaged system, there are three main models you can create:

- **Centralised:** All innovation activities are undertaken by a dedicated team or group within the company
- **De-centralised:** Innovation experts are spread around the various sections of business
- **Hybrid:** A core team manages and oversees the process but most of the innovation activity happens in the business near to the market place.

The centralised model

This is a common model used in business. Many businesses put innovation management into a centre, hire a technical person to run it and ask them to serve the organisation from this central place. There are benefits to this system: for example, it generates a consistent approach to innovation management and uses standardised tools in the process, while project ownership is clearer as are the reporting lines for managing same. However, this system can be slow to respond to market changes and can get bogged down in processes that stifle creativity. In addition, the innovation process under this system can lose touch with other sections of the business and indeed the market. Centralised teams also typically have strong technical skills, but often lack the ability and mandate to foster engagement with the relevant stakeholders of the organisation. While this model yields a cleanly run, professional system, in extreme circumstances it may become an ivory tower detached from the core business groups.

In summary, the **advantages** of a centralised model include:

- Provides a single homogeneous innovation support centre for interaction with the business and its customers (NPD and R&D are elements or sub-sets of 'innovation')
- Is a flexible structure that can be transferred in part or totally into the business at a later stage
- In terms of system control and cost, benefit tracking is less complicated
- Prevents formation of silos (all projects have a technical and commercial element)
- Enables allocation of resources to where they are required (resources follows potential)
- Better optics and better understanding in the outside world
- Provides speedier management decisions
- Better management governance
- Single transparent budget
- Provides for technical and marketing considerations in all areas.

However, the **disadvantages** of a centralised model include:

- It can often result in an 'innovation silo' where innovation is seen as belonging to someone else
- It can become an inflexible bureaucratic structure
- The resources are not in the control of the people at the customer 'coalface' so there is often poor buy-in from key customer-facing staff
- Market insights are difficult to obtain or are not passed on to the centre.

Figure 12: The centralised model

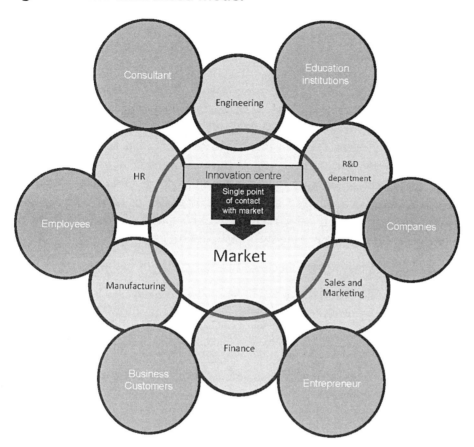

The de-centralised model

This is a less common model, but one that appears to be getting more traction. This model sees multiple centres of leadership set up with key innovation roles dispersed throughout the company. It is often somewhat random, with some departments running high-performing silos while others seem to struggle to get the attention of their management team or indeed from the scaled-down central team. The key disadvantages of this model are that it can create a competitive, rather than a collaborative, culture between groups and can duplicate effort. There is often a lack of focus, with too much emphasis on incremental innovation activities as the business units often are judged on short-term financial metrics. Furthermore, there is a danger of a lack of ownership (as reporting lines are often blurred). And finally, turf wars can – and often do – develop between departments when this model is installed.

In summary, the **advantages** of a de-centralised model include:

- It reduces the risk of the innovation process becoming siloised and removed from the commercial staff at the coal-face interacting with customers
- It is more conducive to an open innovation model
- It increases the touch points of the process with the market place and is therefore more effective at generating market insights than a centralised model
- More staff in your business are involved in the process
- It is arguably a less bureaucratic process for participants and therefore is potentially more spontaneous and creative
- It is less managed and bureaucratic in nature.

However, the **disadvantages** of a de-centralised model include:

- It can result in a lack of focus, as each business unit or section is doing its own thing
- Duplication of effort can occur, with attendant budgetary implications
- It is less conducive to managing external relationships (open innovation)
- Innovation may well take a back seat to other more pressing tasks and only occur when resource availability permits – the 'rainy day' approach
- Cross-business unit opportunities can be missed.

Figure 13: The de-centralised model

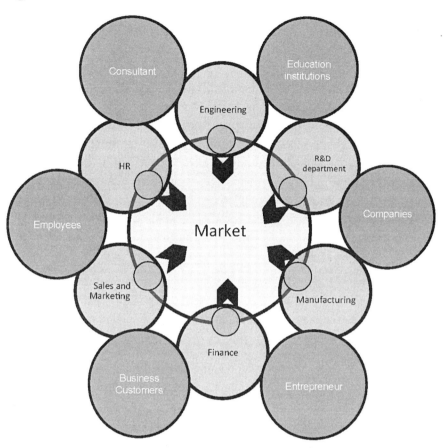

The hybrid model

This is the model you should aspire to establishing – but only when the time is right. It is definitely the most progressive and the most conducive to producing continuous innovation at the pace required. In this model, different business units continue to build their own capacity based on their specific needs. A central core team is established, which directs the whole system toward long-term strategic goals. With this model, the culture of the whole business is more focused and open leadership, market insights and collaboration can be more effectively managed. Often the centre retains activities such as ideas management but a key difference of this model over others is that the central innovation function is a recording and motivating unit rather than being responsible for project delivery. It is, however, often necessary to operate the de-centralised model before this one is put in place in order to establish and mature the competency of the team involved.

It is therefore argued that a de-centralised hybrid model of innovation management is best conducted once the correct people are in place, a system for ideas generation is established and a process for managing same to a system similar to a review gate approach is used.

In summary, the **advantages** of a hybrid model include:

- It may be necessary initially to centralise the innovation function and to recruit / develop the skills needed to implement a robust cross-company system for innovation management. However, once these systems are developed and embedded in the businesses, the most effective means of extracting maximum commercial successes is to transfer some of the resources in central innovation as near as possible to the market place. The hybrid model facilitates this

- A core central team should remain in the centre to identify and capitalise on cross-business unit opportunities and to maintain focus on the external partnerships needed to achieve the full potential of a business transformation objective. Generic functions, such as ideas management, therefore can remain central in order to ensure that they are done! In addition, by identifying cross-business unit opportunities the hybrid model prevents the formation of unit silos

- The portfolio of innovation projects conducted throughout the business can be conducted as close as possible to the market place while market intelligence can recorded at the centre for all to see

- A centre of excellence can be established where the acquisition of grant funding and tax credits can be managed, thereby avoiding unnecessary duplication of effort
- The much scaled-down centre in the hybrid model can be a single point of contact with the outside world in an open innovation model.

Figure 14: The hybrid model

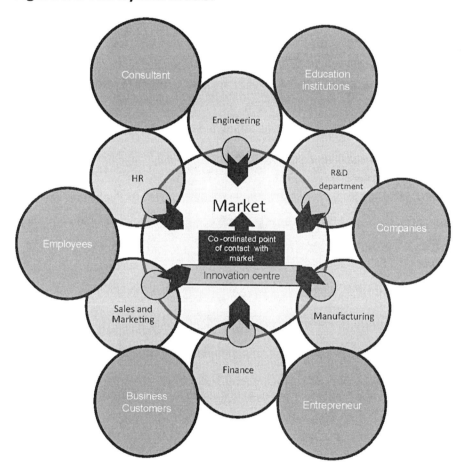

There are no, or at least very few, disadvantages of a hybrid system in my mind. However, be careful that the hybrid model does not gravitate or drift to a de-centralised approach or, worse again, to an informal model. The centre needs to be informed of what is happening in the de-centralised areas and be actively involved in the on-going process of ideas generation, ideas prioritisation, project initiation, project management / reporting and most importantly, the portfolio management process.

As this is the model I favour, the following are additional objectives and functions of this approach to innovation management that I have included in my workplace.

The primary objectives of a hybrid innovation core management team include:

- To maintain a formalised innovation management process (based on market insights)
- To facilitate and track all innovation activities in the business
- To bring new products and services outputs from the process to the market via the business units
- To obtain the maximum grant funding and tax credits for all innovation activities conducted.

The primary functions of a hybrid innovation management team include:

- Maintenance and auditing of an innovation management process across the business
- Co-ordination of ideas generation and prioritisation in the business
- Overseeing and ensuring balance of efforts (portfolio management)
- Setting of innovation strategy to align with the overall vision on the business
- Funding co-ordination
- Communication
- Central repository for all innovation matters in the company
- Manage and develop / expand the business innovation ecosystem.

When the hybrid model is adopted, a number of key groups or participants in the process are identified. I like to call the entire group the 'innovation ecosystem', which, in the modern way of doing things, also includes people and groups outside your business (open innovation).

> **In summary, you need to choose the structure that best suits your business. There is no right or wrong structure. The choice is totally yours. One model does not suit all situations.**

Setting goals, objectives and a dynamic mission for your team

Now that you have chosen a structure for your business, it is necessary to set objectives for the innovation process which must be clearly aligned to the overall vision, goals and resulting strategy of the business.

The core team is the driver of the process. It is vital to set clear goals for them, which must be Specific, Measurable, Achievable, Relevant and Time-bound (SMART) and fully aligned with the overall business goals and objectives.

Your initial objectives may include both primary and supporting objectives – for example:

- Primary objective:
  - To enhance the existing innovation culture company-wide
- Supporting objectives:
  - To introduce structures and processes for the management of the innovation project throughout the company
  - To co-ordinate the recording of the resources employed in innovation / R&D activities group-wide in order to recover the maximum tax credits available and secure maximum grant funding for all spending
  - To introduce initiatives that will promote the development of coherent innovation behaviour group-wide
  - To carry out innovation activities, where possible, in conjunction with research institutions and / or industrial partners, in areas where future potential is identified.

These objectives need to be revisited on a regular basis (for example, annually) as the process matures and the goals change accordingly.

When the process is becoming routine to users, a more mature set of objectives may be considered:

- Primary objective:
  - Embed the innovation process internally and expand externally
- Supporting objective:
  - Drive innovation by feeding ideas into the pipeline, starting projects and delivering new products / services
  - Generate sound market / consumer insights

- o Demonstrate value to stakeholders (internal and external)
- o Create and monitor innovation behaviour
- o Strengthen existing, and develop new, external collaborative networks.

A typical innovation mission statement for the core team might read:

The team will lead, drive and facilitate innovation activities across the business by capturing and enabling new and creative thinking thereby aiding the transformation away from the business's existing activities to areas that are more economically and environmentally sustainable. We will be guided in all our efforts by business needs and grounded by market insights to bring value to the organisation.

We can achieve this because of our expert capability in innovation management, breadth of knowledge to exploit synergies across business sectors and through our ability to engage with internal and external stakeholders.

Building an innovation culture

People often ask me how they can develop an innovation culture in their business. My answer is simple: "Rome was not built in a day". To instil a culture of innovation, you first must introduce innovation behaviour – and for that, you need the correct processes in place.

Therefore, in summary: adherence to a process for innovation in a business leads to innovation-centric behaviour in staff and, eventually, when it becomes embedded in a business, to an innovation culture.

Figure 15: Generating an innovation culture in your organisation

Process ➡ **Behaviour** ➡ **Culture**

To succeed in any innovation activity, an understanding of both the market and the technical problems facing the business are needed. Innovation is not the domain of any one section or group in the business. It is truly a multidisciplinary task and must be resourced accordingly. I like the simplified team-building process produced by tools such as Belbin's team-building model (see Meredith Belbin's 1981 book, Management Teams). The types of individuals on the team are arguably more important than numbers but, in an SME, at least three should be on the team – in larger companies, this number can be six or more.

It is vital that you select a team based on their willingness to be involved. To this end, it is worth remembering that people engage in innovation activities because of a wide range of reasons, including *inter alia*:

- Ambition:
  - They do it to get noticed in a business (from management and staff) in order to advance their career
  - They have a clear vision of where they see the company and themselves going and these are fully aligned with the overall business strategy and they want to get there as fast as possible
- Drive:
  - They have a passion about the activity and are driven to 'get on with it', regardless of obstacles encountered
  - They have great self-belief and a willingness to succeed

- Reward:
    - They want to share in the financial returns of their endeavours
- Recognition:
    - They believe they are, and want to be seen by others as, team players and as good communicators
    - They want to be recognised in the filing of intellectual property (patents, trademarks, etc).

Comment: What structure did we use?

Company – Horseware, Dundalk

Employees – Over 100 are employed in Ireland; an additional 350 are employed worldwide (China, Cambodia, USA)

Sector – Equine welfare

Comment by – Ciaran Herr, Purchasing director

Horseware, I contend, is one of the real success stories of Irish industry. We are now a world leader in horse rug sales, exporting to over 100 countries worldwide. Our premium brand, Rambo, is well recognised and respected in all markets but we also manufacture and sell lower cost items. We are indeed proud of our Irish origins and our humble beginnings and attribute much of our successes to our on-going innovations based on feedback from the market.

The founder / owner Tom Mac Guinness is still very much involved in the business and is the driving force for us innovating, thereby progressing into new markets and introducing new products, with innovations far removed from the ones we started with. It has been through requesting constant feedback from our customers and our intimate knowledge of the market place that we have been able to target our innovation activities so as to achieve the maximum results. We sometimes get frustrated by the widespread belief that it is just the high-tech companies in the pharma or ICT sector that innovate. Innovation is very important to us, so much so that it is in our organisation's DNA. We live and breathe product / service continuous improvement and are looking for the next opportunity for us to develop into – for example, we now consider ourselves to be in the equine welfare business as opposed to horse rug supply only.

For many years now, we have improved our offerings to the market place, constantly changing our products based on market insights. The CEO was the main innovator and discussed his NPD ideas with others, generally over a coffee. It was run on an informal basis but always happened. As we got bigger, however, we have operated a more structured approach to innovation. The need to record time spent on innovation / NPD / R&D activities for grant funding support from Enterprise Ireland or for tax credits has forced us to do this. There is currently a design office with four full-time employees. In the main, most of our innovation activities happen there.

We operate a very consultative approach to our innovation process nowadays (I think 'open innovation' is the term used to describe it). We are not precious about whether something was invented here or not! As we move into new markets with non-core products, we are depending more and more on external sources (third level colleges, private entrepreneurs or other companies wanting a route to market).

In hindsight, we have moved from a closed to an open innovation approach and from an informal structure to one that is more organised in a single office. Formal innovation meetings are held and recorded. However, it is still a process that is centralised around a small number of individuals. Our next step in our innovation evolution is to broaden this process to involve more people.

> **All successful innovators are driven by motivation! So keep the atmosphere dynamic. Assemble a multidisciplinary core team (depending on the structural model selected) that represents all sections in your business. The ownership of projects must remain in the domain of the business functions. Central innovation should be an oversight function, not necessarily a delivery function.**

5: WHERE and WHEN TO INNOVATE

There are no old roads to new directions. (Boston Consulting Group)

The three horizon model, which I use, is a good way to decide where to focus on in an IMS.

Figure 16: The three horizons of innovation model

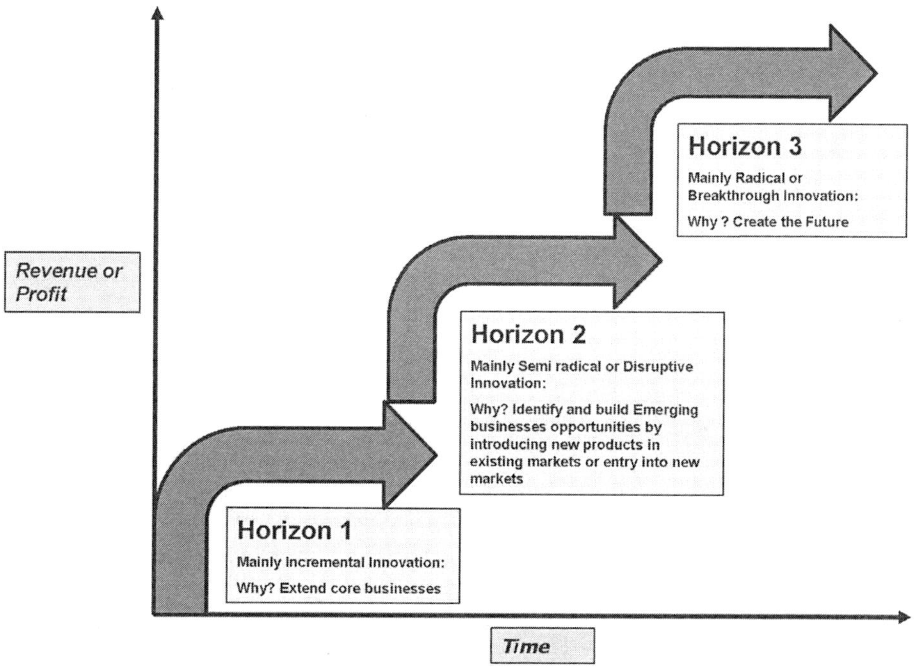

The concept of innovation management across the three horizons was first described by Steve Coley at McKinsey. It has been used and quoted widely ever since, in order to try to convince businesses that the future is important too and that to expend all one's efforts on the here and now is dangerous.

Essentially, the horizon approach describes the 'where' of innovation. Specifically, it identifies where to place your focus in order to maximise returns, while at the same time keeping a firm eye on the 'here and now' of your enterprise (maintaining a balanced portfolio). Essentially, there are three periods in which any business should conduct its innovation activities:

- **Horizon 1 (H1):** The present day innovation activities – the cash cows: Current, cash-generative business activity within a business. Any innovation here is about doing what you do better or making incremental improvements in your current product / service offering

- **Horizon 2 (H2):** The immediate future – the expanders: New business activities that will enable your business to expand past the next number of years and become cash-generative in those activities to enable further innovation. Any innovation here is focused on step change improvements. It is sometimes called semi-radical or disruptive innovation

- **Horizon 3 (H3):** Looking into the future – the futurists: Future business activities or 'tomorrow's world' that will shape the business. Often called 'blue skies' innovation, or radical or breakthrough innovation.

Overly focusing on H1 can result in missing some radical or breakthrough opportunities as a result of which your organisation can be put out of business very quickly by someone else launching a game-changing innovation into your market. On the other hand, too much of a focus on the future may mean that your eye is being taken off your existing business (the cash cow) and you are in danger of being out-competed in the current market.

In any business, innovation activities need to be conducted across the three horizons simultaneously as summarised in the table above.

There is no doubt that a business should focus some of its innovation efforts across all identified horizons. The percentage of time that you should focus on each horizon differs from business to business. Google, for example, uses a 70/20/10 split – most of its innovation resources are put into improving what it is doing now, with proportionally less focus on activities in H2 and H3. *Harvard Business Review* reports that this split is typical and that firms with similar splits outperform the market. Having a 'balanced' approach to innovation does not mean that you must innovate in all three horizons simultaneously in equal percentages; on the contrary, a bias to H1 innovation is necessary.

Table 3: The features of each innovation horizon

Feature	Horizon 1	Horizon 2	Horizon 3
Time focus	**Here and now**	**Immediate future**	**Tomorrow's world**
Duration	Now – 1 year	2 – 5 years	5 years plus
Main focus	Incremental innovation	Semi-radical or disruptive innovation	Radical or breakthrough innovation (blue skies)
Main activity	Just do it (JDI)	New product / services or entry into adjacent markets	Mainly R&D
Examples	Continuous improvement Reduction of costs Avoidance of waste	New products or services Business model innovation	Concept proving Pilot development IP management (patents / licensing)
Main driver	Market-led	Market-led / market-aware	Market-aware
Main objective	Defend and expand core business – look after the cash cow	Identify and develop new business opportunities by introducing new products / services in an existing market, develop new markets for current offerings or identify new ways of doing business	Develop the future

There are multiple ways a business can progress by innovation through the horizons. There are, however, two main routes:

- **Continuous innovation:** Where a business is constantly evolving
- **Continual innovation:** Where discrete advances are made at defined stages of a company's evolution.

In the latter situation, little or no evolution or growth occurs between the stages. This is shown graphically in **Figure 17**.

Figure 17: Continuous and continual innovation through the horizons

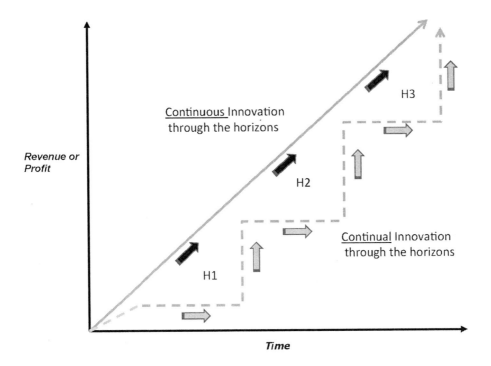

Portfolio management

There should be a happy balance in your business between 'here and now' innovation and 'tomorrow's world'. Too much of the former and you are putting your business in danger of being made redundant in the market place by a breakthrough innovation market entry (ask the Sony Walkman people about this!); too much of the latter and you can become fixated with 'navel-gazing', thereby losing focus on your 'cash cow' and so you go out of business because of poor cash flow.

There is a temptation for the technical people in your organisation to gravitate to the complex innovations of the future. On the other hand, operational and sales people like to focus on the results in the month gone by which are really not very relevant as these sales are gone. The correct ratio of activity across the three horizons will vary depending on:

- Maturity of your existing market and its rate of change
- Competitors (presence or absence, activity level, etc)
- Your presence in the market place (dominant or small player)
- Resources available (people and financial)
- Company strategy and ambition.

There is no fixed formula for what percentage of your effort / resources should be committed related to activities in each horizon. However, it is vital that the business operates a balanced portfolio approach and that the bank of projects currently underway is regularly reviewed to ensure that the division of effort is appropriate to where the market is going.

No textbook on innovation would be complete without a four-quadrant diagram! **Figure 18** is a risk : reward depiction of how a business might manage its portfolio of innovation projects to ensure that scarce resources are properly employed, that dead projects are terminated and that a range of projects in the risk : reward spectrum are progressed concurrently. You need to watch out for developments in Quadrant 3 and be proactive in looking for options here. However, you also need to keep a keen eye on your core business activities and innovate to defend and, if possible, extend your market share here (Quadrant 1). If projects identified as belonging in Quadrant 4 happen, that is great but you should never spend too much of your precious time dreaming. But if they come along grab them! And, of course, projects that are classified as being in Quadrant 2 need to be binned ASAP!

Figure 18: The risk : reward portfolio map

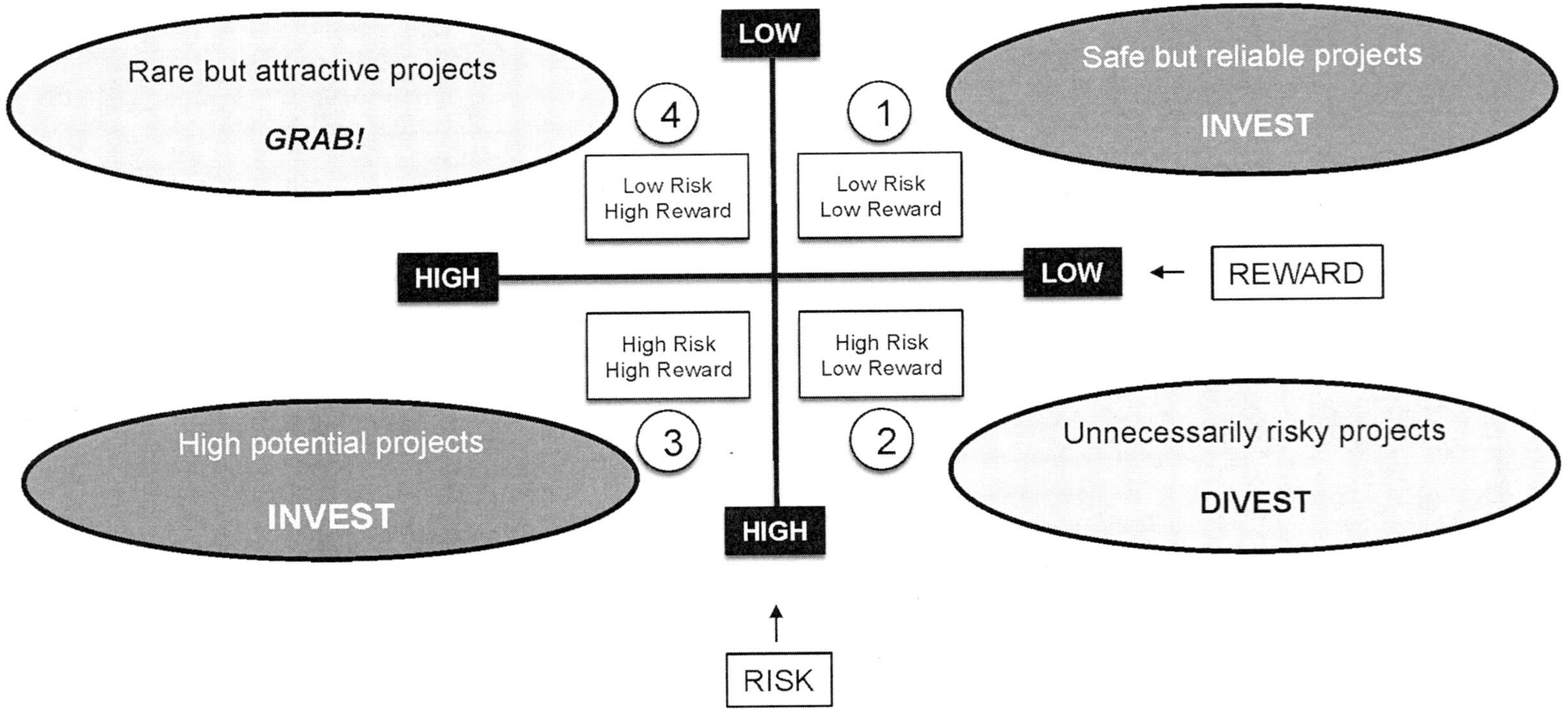

Safe but reliable projects

INVEST

Unnecessarily risky projects

DIVEST

REWARD

LOW

1 Low Risk
   Low Reward

2 High Risk
   Low Reward

LOW

HIGH

RISK

4 Low Risk
   High Reward

3 High Risk
   High Reward

HIGH

Rare but attractive projects

GRAB!

High potential projects

INVEST

The whole idea is to look at reinventing your business in response to market changes (H1 Innovation or Quadrant 1) or to be the radical influencer of change (H3 Innovation or Quadrant 3).

As was said before, innovation is a pseudonym for transformation or change. For example, consider two very different companies that have embraced this change culture and effectively reinvented themselves.

First, an iconic Irish national company, Bord na Móna, which has evolved over the years from a fairly low-tech peat-mining company to one that is much more diversified and more prepared to meet the challenges of the changing regulatory environment that could render its original business model obsolete and possibly challenge its very existence (climate change and carbon tax).

A second, international, example is Nokia, the mobile telephone company. Not many people know that it started life in the mid-1800s in Norway as a wood pulp and later a rubber boots manufacturing company! Through a series of transformations, the company morphed into the world-class technology company that it is today. But even successful companies like Nokia are now coming under pressure in the ever-changing smart phone / mobile phone market. It will have to reinvent itself again and again to survive.

> There is a happy balance between 'here and now' and 'tomorrow's world' innovation for your business. Too much of the former and you are putting your business in danger of being made redundant in the market place by a breakthrough innovation market entry. Too much of the latter and you can become fixated with 'navel-gazing', thereby losing focus on your 'cash cow' and so you go out of business because of poor cash flow.

Comment: Getting the balance right

Company – FBD Insurance, Dublin

Sector – Insurance

Comment by – Catherine Wheelan, HR Manager, FBD Insurance

I am a great believer in operating a balanced approach that addresses the near-term *vs* longer-term aspects of innovation effort but also makes us look differently at operational areas of our business where the greatest return on our efforts can be achieved.

When I was working with insurance group Axa and acting as innovation manager there (as part of my people management or HR role), I looked at all of the innovative ideas that were submitted to the ideas acquisition process over the previous two years, as a result of our innovation drive we called Mad-House. On analysing the ideas submitted that went on to be successfully implemented, I found the following breakdown:

- 10% of ideas were entirely novel
- 10% of ideas were borrowed from an existing idea and implemented in a new application
- 40% of ideas involved continuous improvements
- 40% of ideas centred on eliminating waste.

It is always wise to look at your Innovation efforts based on the short-term or incremental and long-term radical innovation. From the chart below, we can assume that continuous improvement and avoidance of waste is categorised as short-term or incremental innovation while borrowed and new ideas are the longer-term radical options. Therefore, a ratio of 80:20 is achieved.

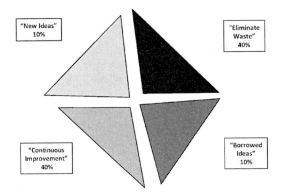

In my career to date, we have always significantly skewed our innovation efforts in the favour of the short-term incremental projects. I think this is vital to retain the involvement of the majority of employees and, just as importantly, to maintain the credibility of the process.

6: IDEAS MANAGEMENT

It is better to have enough ideas for some of them to be wrong, than
to be always right by having no ideas at all. (Edward de Bono)

Sourcing ideas

Ideas are the lifeblood of innovation. Your employees are the first port of
call for an ideas stream. First, let people know what all this innovation talk
is about by circulating an information brochure to all employees. Tell them
you need their assistance for it all to work and set out in plain language
how the system works (such as the information booklet – see later).

The innovation model that my colleagues and I have operated for the past
six years is an open innovation process, initiated through ideas generated
from employees, external sources and collaborations with third parties
such as universities and government agencies. Ideas are captured,
managed and prioritised in line with the company's strategic objectives as
well as meeting existing and future market needs.

'Winning' ideas are then managed through a stage and gate process, a risk
management tool that critiques, evaluates and develops an idea from
discovery to potential commercial development. Following completion of
essential tasks and deliverables within each stage, the project is reviewed
at strategically-placed gates or review points. The gates are quality control
mechanisms between stages that can open to allow a project to move
forward into the next stage or conversely can close to kill the project,
thereby allowing the leveraging of resources into potentially higher value
projects.

When looking for ideas to put through your process, consider these four
main sources:

- **Internal Person ideas** (usually from the employee or customers)
- **Internal Group ideas** (usually from ideation sessions or group thinks)
- **External Person ideas** (usually from an entrepreneur)

- **External Group ideas** (usually via a business or an educational institution).

Internal person-derived ideas (the employee)

Generally, these are incremental ideas involving doing something better or the avoidance of waste. In order to boost the number of ideas in the system, it is strongly recommended that all your employees are encouraged log onto your company's website and / or its intranet and to navigate their way to the innovation section, where there should be some innovation management content. The better sites I have seen included a video clip of the innovation process within the business, testimonials from people who had engaged in the process and an interactive ideas submission facility to encourage people to engage with the innovation team and to 'help shape our immediate future' by submitting ideas.

Internal group-derived ideas (ideation sessions)

Ideation sessions are an efficient means by which to generate ideas. In our case, we calculated that in the region of 15% of the ideas in our 'ideas bank' came from ideation sessions. However, ideation is best used for a defined problem as opposed to a large 'brainstorming' session where there often is a lack of focus.

I have heard it said that to get a good idea, you need to process as many ideas as possible. I believe this is only partially true. It is not entirely a numbers game: the quality of ideas is also important.

A typical attrition from ideas mined to those that overcome the coarse filter (see later) to enter the ideas bank is 50%. It is said that less than 1% of ideas that enter a system (post-coarse filtering) make it to market.

External person-derived ideas (the entrepreneur)

Ideas from this source are generally disruptive or radical ideas and involve doing something new. The expectation of the idea generator is that your business will provide their innovation with a route to the market and they are willing to share the proceeds with you in order to achieve this. Do not be too precious about 'it must be invented here'. It is often an option to license an innovation with an agreed royalty payment based on minimum sales; in addition, an upfront payment to the entrepreneur is sometimes requested to cover patenting and initial development costs. However, in the long term, a business can save a lot of time and money by licensing as opposed to reinventing the wheel!

External group-derived ideas (businesses or third level institutions)

Ideas from these sources are very useful and it pays to engage with them as third level research institutions generally have the resources to do research where the outcome is less certain – for example, higher risk or radical innovation. Most institutions now operate a Technology Transfer Office (which sounds as if it only applies to technology products, but this is not the case). It is well worth your time taking a trip around the relevant third level research institutions, getting to know what they are or have been working on and investigating whether your business can provide a route to market for any of their output. Ideas from this source can come from your direct contacts with the institutions or from the external ideas mining machinery you have in place (your internet portal). Direct business to business (B2B) ideas sourcing by licencing, partnering or M&A is another route.

Whether they are internally or externally-derived ideas, from an individual or a group, **Table 4** shows some of the places to go for the best results and the type of ideas you might expect from each source.

Table 4: Ideas sourcing

IDEA SOURCE	TYPE OF IDEA (HORIZON)	MAINLY FROM WHERE?	MAINLY FROM WHOM?
Ideation sessions	1,2 or 3	Internal	Groups
Company intranet	1 or 2	Internal	Individual
Suggestion boxes	1 or 2	Internal	Individual
Third level TTOs	1,2 or 3	External	Groups
Conferences / seminars	1,2 or 3	External	Both
Internet	2 or 3	External	Both
Competitions	2 or 3	Both	Both
Mailshots	1 or 2	Both	Both
Innovation seminars or 'fairs'	1, 2 or 3	Both	Both

The intranet

This has proven a useful way of getting the message out to employees. However, it has proven to be a sparse provider of ideas. On speaking with many of the users of the process, it was almost unanimous that an intranet is rarely used as a means of keeping up-to-date on innovation activities in a business or submitting an idea to the process. On reflection, it may be necessary to have some type of 'bait' to attract users onto the site and then to direct them to the innovation ideas section. The use of a log in portal system, where employees access their systems through the intranet, is an option worth considering. In any event, the employee intranet source of ideas generally provided ideas of incremental or H1 nature and rarely more radical or H3 ones.

Figure 19: Screenshot showing an intranet innovation page with an ideas submission icon

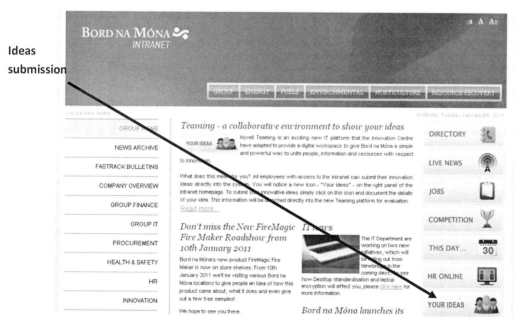

The Internet

This is truly an open innovation ideas generation tool as it is available to anyone inside or outside the business.

It is generally a good means of attracting a lot of ideas from the general public, without having to resort to expensive advertising or holding time-consuming meetings. It also provides a means by which a business can become involved in crowd-sourcing. Ideas from this source are generally in the disruptive or radical territory but suggestions on how to improve the current market offering also can be acquired, especially from consumers. The Internet portal needs to be kept up-to-date and regularly refreshed so as to attract return traffic.

Figure 20: Screenshot showing an Internet innovation page with an ideas submission icon

Letter to the management team in your business and their direct reports

Again, in Bord na Móna, we have found this to be a useful and relatively inexpensive means of disseminating information to potential partners and employees who could form part of our innovation ecosystem.

To: The Management Team

Re: Launching the Innovation Ideas Generation Process

Colleagues,

We are pleased to announce the official launch of the Ideas Generation element of the Innovation Initiative at Blogs Ltd. whereby anyone in the company can submit business ideas for formal assessment. If successful, ideas will be progressed into commercially successful business ventures. Ideas are the pipeline from which we will generate a successful innovation campaign. They are the lifeblood of the process and without a steady flow of them the entire process eventually will grind to a halt.

All line managers in the company need to be aware of how the entire innovation process is intended to work, from the ideas generation, through screening processes, to the selection of which projects to run with. These projects then will be implemented and, where successful, commercialised. To this end, I enclose a booklet summarising the process. The booklet outlines how the process works and provides you with enough information to actively participate in the generation and selection of ideas. In the short term, it will assist you in dealing with any queries that you might receive from your teams.

As a key member of the leadership group in the company, you can play a critical role in the process by assisting in the generation and capturing of ideas and submitting them into the system through your innovation satellite team. Your help also will be invaluable in playing an enthusiastic and committed role in the operation of the scheme.

If we were asked to select just two main messages that we need you to be aware of and to convey to all staff in the company (and indeed beyond) with whom you come into contact, they would be:

- **Innovation is for everybody.** It is as much as about the factory worker as it is about senior executives in the company. It is not just the domain of a few 'techies' in the centre. It must be embedded throughout the company and beyond in order to reach its full potential

- **Innovation is not just about 'blue skies'** (the flashing light bulb!).
  Futuristic radical innovation plays a secondary role to the incremental
  'here and now' continuous improvement and cost savings.

I need your help to assist in the successful embedding of the innovation
process company-wide and getting these key messages out to all. I know we
can rely on you to make this a successful process and to help transform our
company to provide a sustainable future for us all.

Thank you for your support.

Regards

Mr Smith

Innovation management co-ordinator

Employee poster campaign

This is a means of extracting ideas from your staff and direct customers
without them having to deal with management directly on the matter. It
usually attracts H1 ideas, as employees are generally concerned with the
'here and now' but occasionally a H3 idea can arise from the initiative also.

Figure 21: Employee poster requesting ideas submission

Ideation sessions

These can be a useful source of ideas. In the past, I have found them a productive means of ideas mining, generating up to 15% of the ideas logged in any time period but amounting to less than 5% of the innovation oversight manager's time. At the outset, the 'known knowns' are defined before the creative thinking process begins. The meeting or session then leads to the tougher phase of using a divergent / convergent approach to explore all options and to identify solutions. The diagram below summarises this innovation ideation session approach.

Figure 22: Ideation sessions

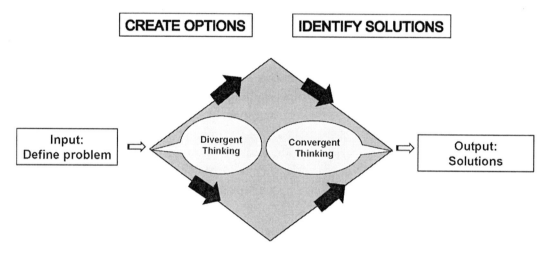

Internal competitions

These may be in the form of a quiz about the innovation process in the company (advertised on the intranet and on noticeboard posters) or a draw for participation in the submission of ideas. The winners of the competitions should be recognised in the company's periodical if there is one (a sort of 'innovation roundup') and / or on the company noticeboards. Some businesses have a scheme for rewarding participants by holding draws for clearly identifiable company innovation management T-shirts that can be worn by the winner as a 'badge of honour'. Some businesses even offer a cash incentive for ideas submitted or projects started.

External competitions

These can be media-advertised competitions or sponsorships or indeed a quiz-like process with a prize. They often attract ideas from the H2 and H3 category. These campaigns can be used primarily to raise awareness of the process but also can be valuable in conjunction with the company website to attract external ideas in the spirit of open innovation.

Third level institutions and their Technology Transfer Offices (TTOs)

Ideas from these sources are very useful and it pays to engage with these sources, as third level research institutions generally have the resources to do research where the outcome is less certain – higher risk or radical innovation. Most institutions now operate a Technology Transfer Office (TTO), which sounds as if it only applies to technology products but this is not the case. It is well worth your time taking a trip around the relevant third level colleges, getting to know what they are or have being working on and investigating if your business can provide a route to market for their outputs. It is vital that a record of whom you met with and what the topic of discussion was so that you can refer to it into the future. It also pays to build up a record of which topics a specific institution has a specific expertise or experience in. You should keep a record of all your meetings with these institutions for future reference as shown in **Table 5**.

Table 5: Recording research institution contacts

Date	Organisation	Who you met with	Topic of discussion	Follow -up

Innovation seminars, fairs or other group events

These are good occasions to showcase the innovators or business transformation activities in your business to other employees and to a wider audience of external stakeholders. They are useful occasions to celebrate the successes of the process, to garner support for the initiative from senior management and to profile your key participants in the process. In truth, they are more to do with awareness generation than ideas generation but, by inviting business leaders as well as technical, marketing, financial, HR staff from your business in addition to as broad a range as possible of external stakeholders, regulators and NGOs, cross-pollination of ideas may arise with attendant opportunity identification.

Hold these events on an infrequent basis (perhaps annually) in order to make the most of the event. Hold the event off-site, away from distractions in the office and limit the use of external communication devices to breaks. Consider the possibility of holding a competition during the seminar where attendees vote or 'virtually' invest in what they consider to be the best prospect in the innovation pipeline exhibited. External facilitation and participation in the event can work well.

What information do you provide to employees and other stakeholders on the process?

In my experience, stakeholders, other than those directly involved in the IMS, require as little information as possible up front but need to have easy access to a reference document when required.

The core team needs more information than the 'family' or community groups and the members of the wider ecosystem need less again. The following is a summary of the type of information you need to provide to all employees and external stakeholders (including customers) in your innovation ecosystem.

INNOVATION MANAGEMENT IN BLOGS Ltd.

Innovation in practice: Securing a future for our business

What is innovation?

Innovation is the harnessing of ideas for the commercial benefit of Blogs Ltd. Such ideas range from continuous improvement and cost reduction to radical new approaches. The aim of innovation is, through these ideas, to drive growth and profitability within the company.

What are innovative ideas?

It might surprise you to hear that 'new' ideas or 'light bulb moments' are only a small portion of innovation ideas. Generally, in business:

- 10% are 'new ideas' or innovation ideas that are entirely novel
- 10% are 'borrowed' or use an existing ideas in a new application
- 40% of ideas involve continuous improvements
- 40% of ideas centre on eliminating waste.

Breakdown of innovative ideas (Cranfield University)

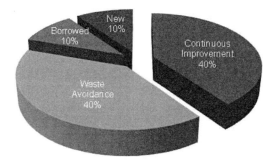

Innovation is for everybody and all ideas are welcome.

How do you submit your idea?

All innovation ideas are submitted on an ideas form. These forms are available at all of the following locations:

- Reception areas
- Production areas
- Stores
- Intranet
- Internet
- Your line manager
- Formal ideation sessions run in the businesses
- The marketing / sales department

How do you fill in the form?

To fill in the form, answer the following questions:

- Is the idea of benefit to our business?
- What are the technical issues and how might these be overcome?
- What is the need for this idea?
- Does the idea fit with our business vision?

Don't worry if you cannot answer all questions, please submit the form anyway.

If you would like assistance or advice in filling in the form, you can ask your supervisor or your Local Innovation Contact.

Who do you give your completed form to?

There are different ways you could submit this form. You can give it to:

- Your supervisor

- Your Local Innovation Contact
- Any member of the Marketing team.

Who is your Local Innovation Contact?

The Local Innovation Contact people are:

Location	Name	Contact Details

What will happen to your idea?

It will be logged and recorded as your idea. Thereafter:

- All ideas will be assessed by the innovation process
- Ideas showing the best potential will be progressed
- Successful ideas (big and small) will be implemented
- Other ideas with potential will be retained and may be implemented later
- Ideas applicable to other parts of the company will be transferred
- Ideas judged not viable will be declined.

What feedback do you get?

You will receive an acknowledgement when your idea has been logged and recorded. You will be informed if your idea is being progressed, implemented or declined.

At any time you can ask your Local Innovation Contact or any member of the Energy Innovation Team about the status of your idea.

What happens if more than one person comes up with the same idea?

Fill in the ideas form as normal, and enter all names in the Idea Creator section. All idea creators will be credited with the idea.

What if my idea has already been submitted?

The person who comes up with the idea first will be credited with the idea, so if you have an idea, act quickly!!

Ideas submission form

Irrespective of the source of an idea, a standardised ideas form on the proposal should be completed.

The form should provide sufficient information to enable the evaluator(s) to process the idea through a coarse assessment process (coarse filter) and therefore provide some assessment of the following:

- Has the idea a fighting chance of making money?
- Is the idea broadly technically feasible?
- Is there a market for the product / service if successfully developed?
- Is the idea aligned with the company vision / strategy?

Figure 23: An ideas submission form

Idea title			
Brief description of idea (What? When? Why? How?)			
Benefit to the company			
Feasibility			
Market need			
Is it aligned with the company strategy?			
Submitted by			
Location		Contact number	
Manager		Area / Business unit	
Received by		Ideas number	
Date		Date by which reply is due	

How are ideas prioritised?

Regardless of the origin of ideas, they all should enter the same process and be prioritised before entering an ideas bank.

The process is summarised as follows:

- All ideas enter the innovation 'funnel' for processing
- A coarse filter is applied by the innovation team in order to weed out some projects
- All surviving ideas enter the ideas bank for scoring
- If a project scores above the hurdle rate, it enters as a live project (the hurdle rate may be changed from time to time)
- All successful ideas enter into the review gate process.

Figure 24: The innovation ideas funnel

As referenced above, the main function of the coarse filter is to separate the dross from ideas that deserve to be considered further. Admittedly, this is a subjective call and sometimes real gems of projects can be missed but, by identifying a number of criteria to assess the raw idea, you can introduce an element of objectivity into the process.

The criteria in this filter may include (Yes or No answers):

- Is the idea broadly technically feasible (can it be done)?
- Is there a market for the product if successfully developed (is there a need for it)?

- Has it a fighting chance of making money (will it sell at a margin acceptable to the business)?
- Is the idea aligned with the company vision / strategy (is it consistent with what we want for our business)?

If the idea passes this test, it moves to a scoring exercise where the central team meets in order to put an arbitrary score on the idea. If it receives a score in excess of the predetermined hurdle rate (which can vary based on the number of post-coarse filter ideas in the system and / or resources to proceed with the idea to project stage), it proceeds to become a project. If it does not pass the coarse filter, it remains behind to be killed or parked!

In a standard scoring matrix, there are typically five criteria adoptable by all sections of the business:

- Strategy
- Technical feasibility
- Cost benefit (financial viability)
- Market need
- Resources.

In an attempt to give the innovation / NPD management more focus, it is prudent to regularly embark on a comprehensive 'sanitising' exercise on ideas in your ideas bank and among projects underway. Some projects may be reclassified as JDI or 'Just Do It' projects and fast-tracked; others may be discontinued or suspended in order to focus on those that show more potential. All of the remaining projects should be recorded on the IMS, where full technical and commercial justifications should be held for them. The overarching aim of this 'sanitising' exercise is to retain a focus on the projects that have the most likelihood of commercial success in the medium-term.

Table 6: An ideas scoring matrix

Rating	10	8	6	0	Weight	Score
Strategic fit	Fits strategy + objectives	Fits strategy	Fits vision	No fit	Business specific	Rating x Weight
Technical feasibility	Can do, within current capability	Can do, by developing new capability	Significant technical risk	High-technical risk	Business specific	Rating x Weight
Cost / benefit	Low cost, high potential, low risk	Medium cost, high potential, some risk	Low cost, medium potential, some risk	High cost, low potential	Business specific	Rating x Weight
Market need	Premium product, meets significant unmet need	New product, meets need with higher margin	Replacing existing product, same margin	Replacing existing product, lower margin	Business specific	Rating x Weight
Resources available	Resources available in-house currently	Resources needed, equal scale to previous project	Significant resources needed, larger scale	Beyond current resource capabilities	Business specific	Rating x Weight

Comment: Which areas did we prioritise and why?

Company – An Post

Employees – Over 9,000 nationwide

Sector – Postal, retail and financial services

Case Study by – John McConnell, Director of Innovation & Quality

From its beginnings in the late 16th and early 17th centuries, the Irish Post Office has grown and adapted to the needs of the times. Telegraph, banking and telephone services were gradually added to mail services. Today, An Post continues to adapt and is now a major commercial semi-state organisation providing a wide range of services that encompass postal, communication, retail and financial services. In fact, it is one of Ireland's largest companies, directly employing over 9,000 people through its national network of retail, processing and delivery points. It also provides agency services for Government Departments, the National Treasury Management Agency (NTMA), Premier Lotteries Ireland Ltd and many other commercial bodies.

The company's core mails business has been in decline for several years now due to e-substitution / digital communication and the challenging business climate among other factors. Therefore the need for innovation is stronger today than ever before. Making innovation a priority of the business has strengthened An Post's position as it faced the challenging years ahead. The company has integrated many of the processes, governance structures and focus areas that are central to driving innovation in any organisation. With these in place, An Post has focused on cost control and competitiveness and also has made strategic investments in key revenue-generating areas such as parcels, retail and financial services and in its group subsidiaries. Thanks to the explosive growth in ecommerce, there are now even more growth opportunities.

The company's priority is to ensure that it continues to monitor consumer trends, as well as its business customers' needs (innovation based on customer insights). Notwithstanding its short-term agenda of cost control and competitiveness, the focus is firmly on the medium to longer-term opportunities for An Post, whilst ensuring that the most effective business model is in place to deliver success into the future.

How ideas are turned into projects

The review or stage-gate model for innovation management is a project management technique (first developed by Cooper in 2001) in which an initiative or project is divided into stages or phases, separated by gates or hurdles. At each gate, the continuation of the process is decided by a manager or a steering committee (often called gatekeepers). The decision to continue a project or not is based on the information available at the time, including the business case, risk analysis, and availability of necessary resources such as money or people. To progress through the gate to the next phase, the gatekeeper must be convinced that there is merit in the project doing so. It is essentially a systematic approach to innovation management that prevents resources being allocated to a particular project until certain criteria are first proven. In simple terms, it is a 'fail quickly and cheaply' insurance system.

The review gate system manages projects from idea to market launch. It is a risk management tool designed to develop the maximum numbers of ideas, with the minimum amount of resources and in the shortest timeframe. It does this by assessing technical, marketing and financial criteria as the project progresses from the initial 'quick look' investigation to full market launch.

The process is shown diagrammatically in **Figures 25** and **26**.

In project review gate management, three (or more commonly five) stages are usually recognised:

- Preliminary investigation (the quick look – Review Gate 1)
- Detailed investigation (the business case – Review Gate 2)
- Development (project management – Review Gate 3)
- Pre-production (the prototype – Review Gate 4)
- Market launch (getting the return – Review Gate 5).

The review gates identified and the requirements of each review are shown diagrammatically in **Figure 27** and in **Table 7** below.

At each review stage from 1 to 5, the level of detail increases and the gatekeeper's level of confidence in their prediction of success or failure of the innovation gets stronger. Remember that the philosophy of this staged approach is to fail early and fail cheaply.

Figure 26: Review gate management

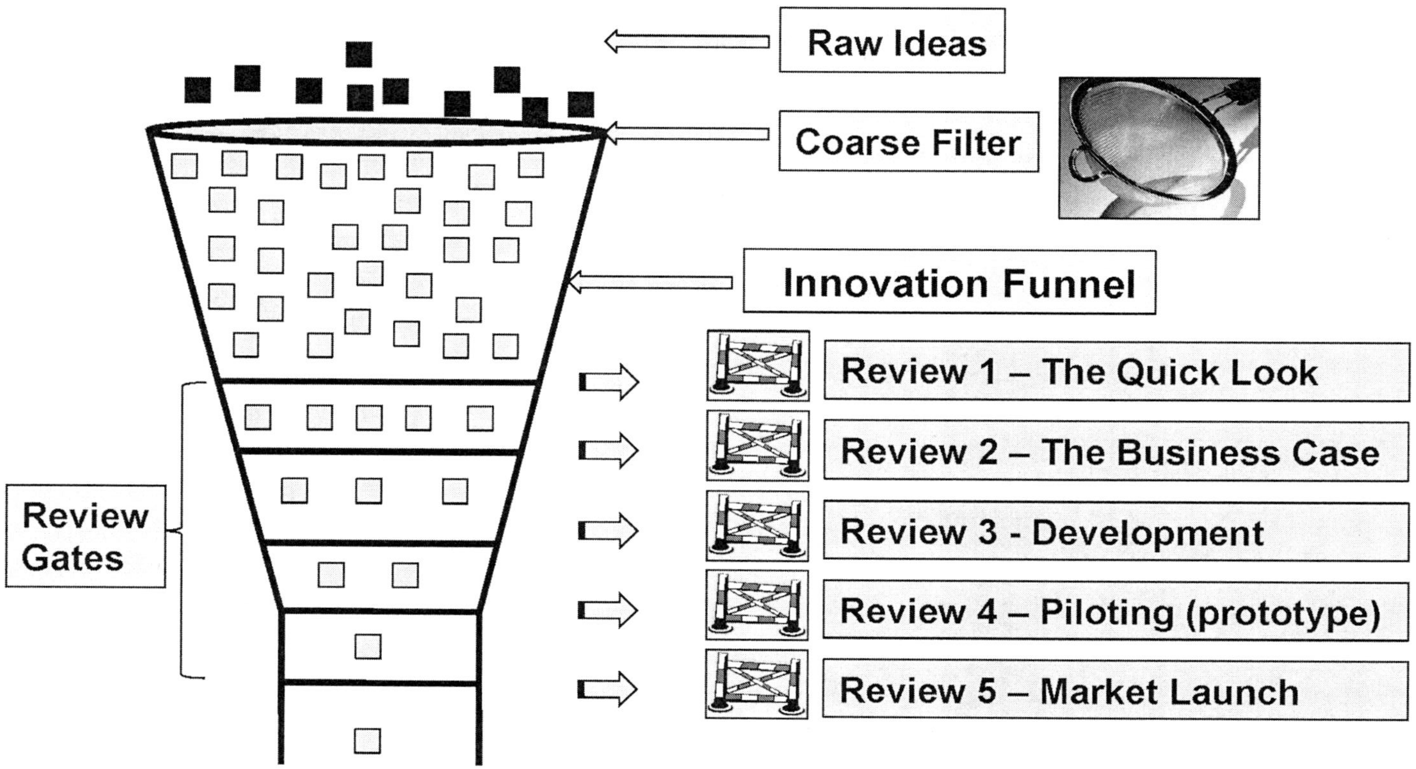

Raw Ideas

Coarse Filter

Innovation Funnel

Review 1 – The Quick Look

Review 2 – The Business Case

Review 3 - Development

Review 4 – Piloting (prototype)

Review 5 – Market Launch

Review Gates

Figure 26: Review gate management

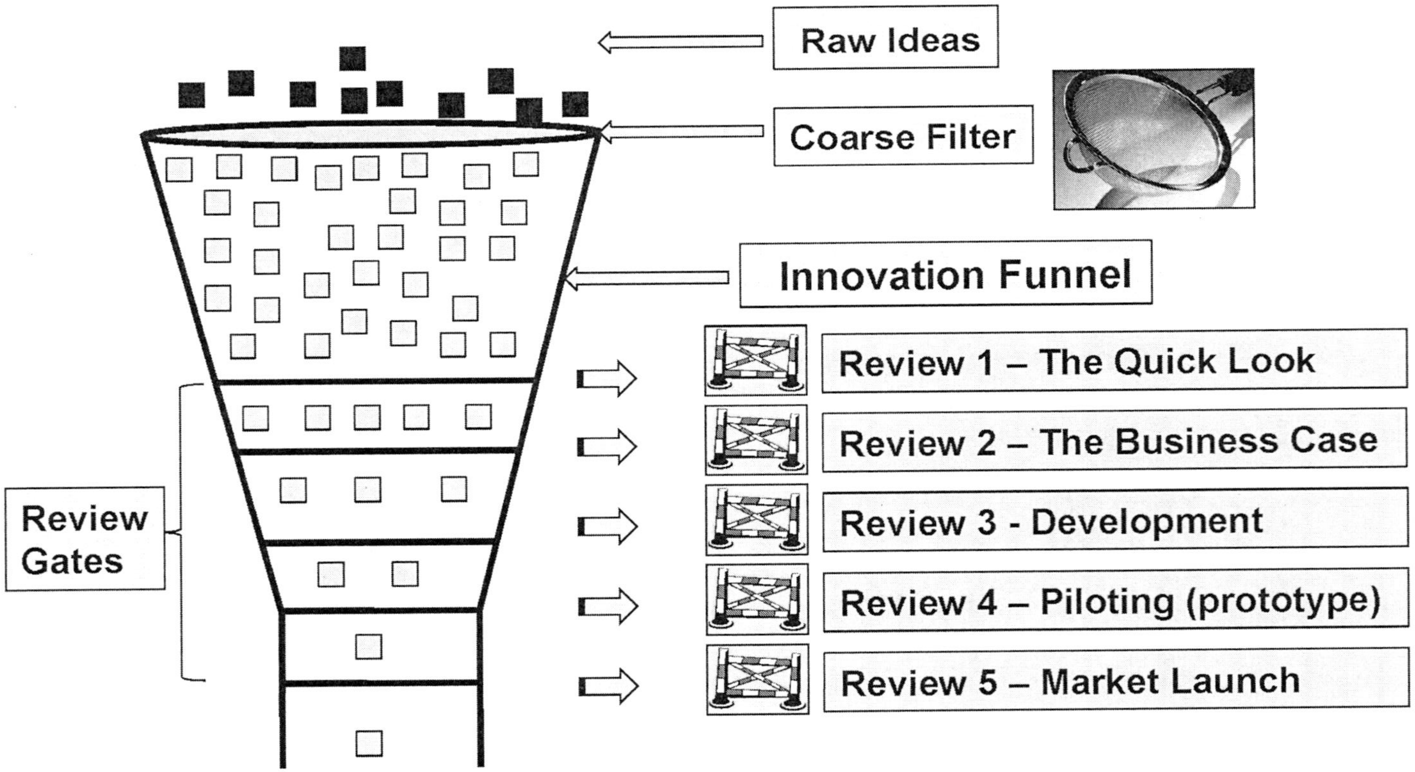

Raw Ideas

Coarse Filter

Innovation Funnel

Review 1 – The Quick Look

Review 2 – The Business Case

Review 3 - Development

Review 4 – Piloting (prototype)

Review 5 – Market Launch

Review Gates

Figure 27: An innovation management funnel with a five-gate review process

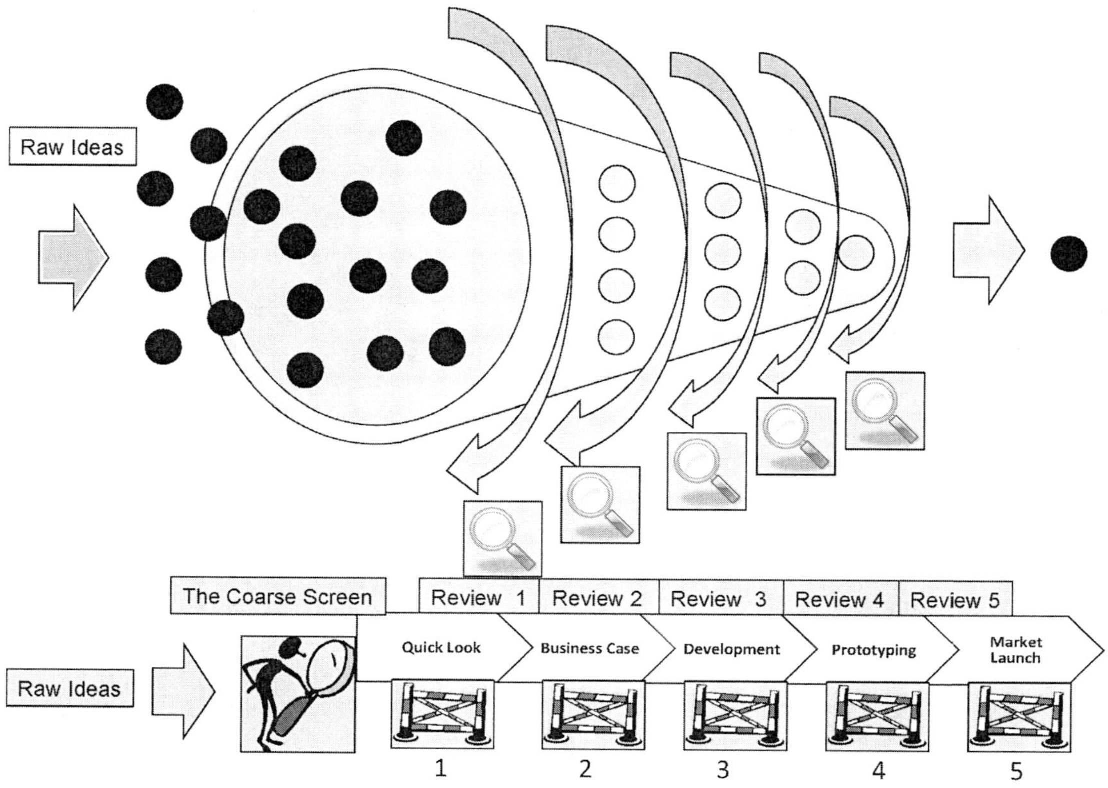

Table 7: The general requirements of each review gate

REVIEW GATE / FEATURE	MARKET INFORMATION REQUIRED	TECHNICAL INFORMATION REQUIRED	FINANCIAL INFORMATION REQUIRED	OUTPUT
1: Quick look	What is the size of the prize? Describe the proposal. What is it? What does it do? Customers and competition: who are they? Who needs and wants it? Quantify market size, share of revenue, etc	Is it possible? Are there any other solutions available? What regulations apply? How does the idea work and perform? High level risk? Uncertainty? Track record? Is there any IP? Infringement?	Is there a return? Compare marketing / technical information available: Return? Costs? Investment? Resources required? Equipment required?	High-level justification
2: Business case	Verify the prize Full competitor analysis: who else is out there? Customer study: what are their needs and wants? Market analysis: size, share of revenue potential	Feasibility outline specifications Standards? Regulations? Licences? Quality? Technical Risk Has it been done before? Uncertainty? Feasibility What are the 'Go' or 'No go' tests? IP Protection Detailed search	Build the model NPV, IRR, ROI, CAPEX, OPEX P&L and cashflow projections? Scenario analysis? Investment costs? Capital required? Resources required? WACC?	Business case and initial financial model

REVIEW GATE / FEATURE	MARKET INFORMATION REQUIRED	TECHNICAL INFORMATION REQUIRED	FINANCIAL INFORMATION REQUIRED	OUTPUT
3: Development	Step 1: Identify options Structured brainstorming; Multiple concepts identified; Develop concepts identified Step 2: Choose the best way to make and commercialise the invention Initial design of preferred option; Optimise the initial design by using appropriate tools / techniques; Sense-check all assumptions Step 3: Final design Sign off on preferred design for implementation; Sourcing of raw materials – Suppliers? Costs?; Modifications – Feedback from user testing? Process Design – Piloting? Full Scale? Step 4: Marketing, compliance and financial Marketing, Compliance and financial update			Business case and detailed financial model
4: Pre-production & validation of market	Step 1: Piloting Set up new plant prototype Step 2: Testing and validation Test and validate design of the plant Step 3: Field trials Conduct manufacturing trials to mimic real life conditions Step 4: Market launch plan Launch plan, Logistics plan, Budget and financial plan preparation Step 5: Compliance All required permissions and permits in place Step 6: Production Full-scale manufacturing process trials; Procurement of appropriate equipment/raw materials etc; Logistics and marketing/sales update Step 7: Financials confirmation Financial model; Risks mitigation process			Ramp-up plan, Launch plan and strong financials
5: Full production & market launch	Pre-Launch Ramp-up all processes, Check documentation, Market launch date set Post-Launch appraisal Technical, Marketing / sales, Manufacturing, Financial, On-going Training, Monitoring and control, Reporting			Handover to Operations and Marketing / Sales and review of Lessons Learned

Innovation attrition rate in the review gates

Figure 28: Typical attrition rates at each review gate

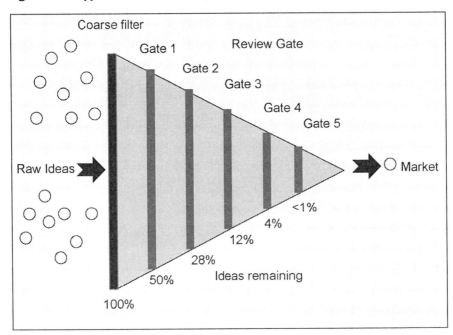

This is simply the rate of project 'die-off' at the defined review gates on the journey through the funnel. It is often said that typically only about 1% of ideas are a market success. But which 1% will it be? You have to kiss a lot of frogs before you identify the prince!

Protecting your ideas

In the past it was customary to value a business in terms of its fixed or 'tangible assets' – the buildings and hardware it owned, the stock it possessed, the vehicles it had on its books, etc. However, in recent years, it is often the case that the non-fixed or 'intangible assets' of a business are just as important to a valuation, if not more important than the 'bricks and mortar'. This is particularly the case for start-ups where the fixed assets of the business have not developed yet. In these cases, IP protection can be a useful means of protecting a business's invention so that it can be free and unrestricted to investigate means by which the invention can be commercialised without the fear that others will copy it.

Therefore, products, or in some cases services, can be protected by a patent, trade mark or design right. The value of IP built up through innovation activities can be realised by the business through the direct selling of an invention in the market place and / or licencing it out to others to sell in markets you are not or do not intend to be involved in. As the licensor, you receive an agreed royalty payment for sales made by the licensee in accordance with an agreed minimum sales volume. A once-off upfront payment to secure exclusivity in the defined market also may be paid by the licensee.

In my experience, IP protection, particularly patents, can act as a very useful deterrent to others who might copy your invention or parts thereof. However, very often a significant additional investment is required in order to defend or enforce the patent. Some companies (particularly start-ups), entrepreneurs and research institutions spend too much of their scarce resources on IP protection issues at an early stage. This is because of an over-focus on technical matters at the risk of losing sight of the commercial issues (remember the difference between invention and innovation referred to in **Chapter 1**). This can be fatal for any good innovation or invention, especially in the funding 'Valley of Death' phase (see **Chapter 9**).

To go into this highly specialised area in more detail is beyond the scope of this book. If, however, you want more detail on IP management, refer to the very comprehensive textbook, **Intellectual Property: From Creation to Commercialisation: A Practical Guide for Innovators & Researchers**, written by John Mc Manus and published by Oak Tree Press. You also should seek independent professional legal and / or other professional advice specific to your own business requirements.

Finally on this topic, it is worth noting that under the Republic of Ireland's tax regime, a number of tax incentives are available to companies and individuals involved in the development and exploitation of IP. In addition, tax credits are available to businesses that carry out R&D under a formal or informal IMS (see **Chapter 9**).

Project tracking

The tracking of ideas / projects as they move through the funnel is critical in order to ensure that a common system is operating throughout the company and that value for money is being achieved. A project tracking system (run on the general business IT platform) should be introduced in order to monitor the progress of all projects underway. The tracking system should include a group collaborative platform.

Figure 29: Screen shot of an IT-enabled innovation collaborative workspace

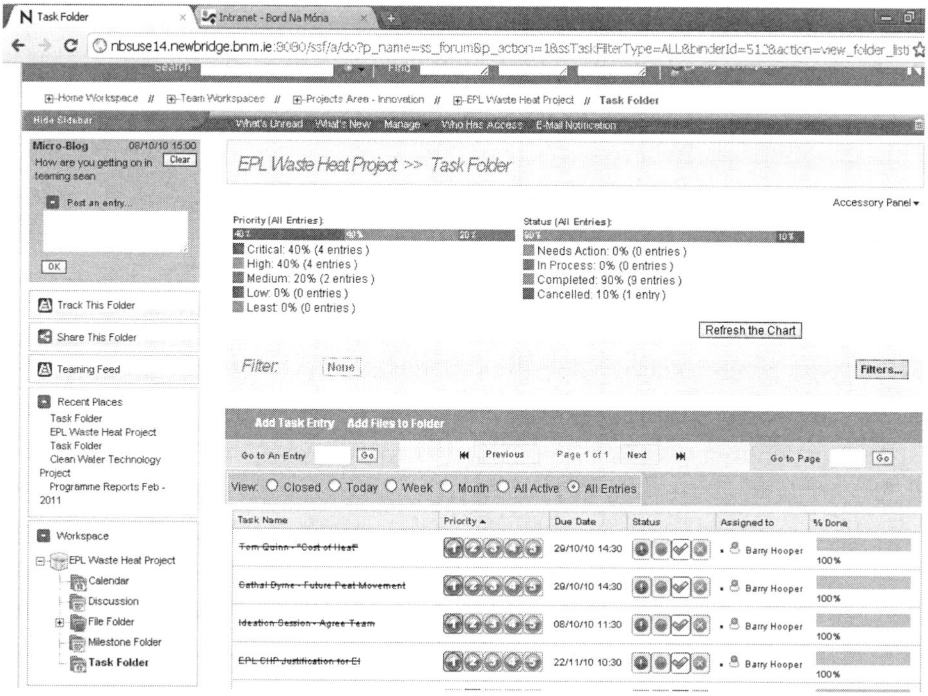

My team and I have installed an off-the-shelf collaborative platform (with some customisation) in my current place of work. This is an IT-enabled platform that was adapted to provide a digital workspace to give the business a simple and powerful way to view and track the progress of projects underway and also for management to view and change its portfolio based on risk : reward criteria. A collaborative platform's main function is to unite people, information and resources with respect to innovation management. The platform has been interfaced with the company intranet, so now all employees with access to the intranet can submit their innovation ideas directly into the system. This collaboration tool provides a digital 'office' where the members of physically dispersed teams can gather online to co-ordinate their efforts and to bring together the right information, resources and skill sets for a project, purpose or topic of interest.

The system should provide details on each active project in the company, as well as track the generation and movement of ideas in the system. The system should record *inter alia* the following detail:

- Number of raw ideas received
- Number of ideas passing through the coarse filter
- Scoring of ideas in ideas bank
- Project title and location (who are the owners?)
- Project sponsor and project manager
- Objective and scope of project
- Category of project (horizon / near-term / longer-term)
- Project status summary
- Project review gate in which the project resides:
    o Preliminary investigation (the quick look)
    o Detailed investigation (the business case)
    o Development (project management)
    o Pre-production (the prototype)
    o Market launch (getting the return).

Any project should require the authorisation of a designated approver ('gatekeeper') before it can be progressed through a review gate into the next stage.

Project management

Once a project is started, good project management (PM) disciplines should be used to record, track and manage its progression. The review gate management system is ideally run as a complementary check on the PM process. It is not a substitute for, or an alternative to, good PM.

At the very least, a project manager should include the following key steps in the process:

- Planning and design phase
- Project initiation
- Project execution
- Project close-off.

For more information on the disciplines of PM in business projects, refer to the Institute of Project Management of Ireland (IPMI).

7: MEASURING PERFORMANCE

Chance favours the prepared mind. (Louis Pasteur)

Measuring performance through tracking key performance indicators

Building on the identified performance metrics above while considering the information available, your IMS should implement management KPIs at three levels.

Level 1 – Engagement and Activity

These KPIs use information currently available from the innovation management process and measure engagement with the process, general progress of ideas through the pipeline and the timeline of the ideas and projects underway (H1 to H3) – for example:

- Number of ideas submitted (total, per business unit, per horizon category)
- Number of projects underway (total, per business unit, per horizon)
- Number of new ideas submitted (total, per business unit per month
- Number of projects underway (by progress through pipeline – investigation to implementation).

Level 2 – Value of Projects in Pipeline

In our case, currently limited financial information is made available on projects. Nonetheless, the following KPIs are proposed:

- Value of projects underway (total, per business unit, per horizon category)
- Potential revenue from projects underway (total, per business unit, per horizon category).

These KPIs inform portfolio decisions on which projects to undertake to maximise value and to support company strategy.

Level 3 – Outcomes of Pipeline

The final group of KPIs stem from the outputs of the innovation process itself, measuring the benefits accruing to the organisation from the innovation investment:

- Number of products launched
- Revenue from new products launched
- EBITDA from new products launched
- ROI / IRR of the innovation management process.

It is critical that an innovation process demonstrates its value to a business. The following KPIs are examples of what could be used to assess the performance of a formalised IMS in a business.

Non-financial metrics include:

- Total number of ideas submitted
- Ideas submitted per group
- Number of patents registered or licences entered into
- Customer satisfaction assessment
- Number of new products launched to the market
- External reputation (is your business a place in which people want to work and the general public considers innovative and progressive?)
- Timelines from idea to market (it should reduce over time) – the innovation cycle
- No or % of ideas progressing to project
- No of projects underway.

Financial metrics include:

- Revenue from new products / services
- Financial value of projects in the innovation system and their potential
- Financial impact: Cost *vs* benefits
- Number of products launched
- Revenue from new products launched
- EBITDA from new products launched
- ROI of innovation management process.

A typical innovation management reporting dashboard is presented in **Figure 30**.

Figure 30: A typical innovation management reporting dashboard

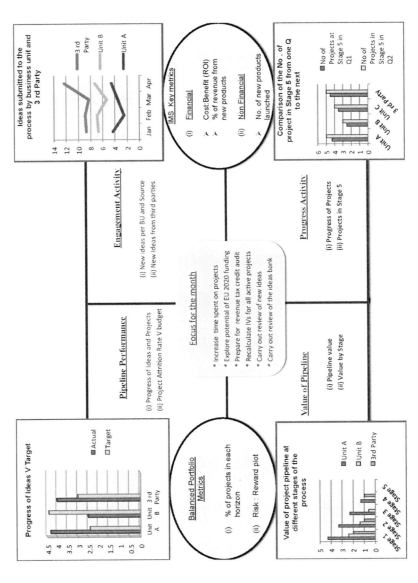

Valuing your innovation pipeline

How do you assess the potential value of your innovation pipeline in order to give management and employees alike an idea of what is coming next (and when) and its revenue-generating expectations? This is a question that has been to the fore of many IMSs since the inception of the process. Failure to put a value on what has been and is being worked on may result in the loss of support for the process from key influencers in your business.

Figure 31: Innovation value – A useful financial metric?

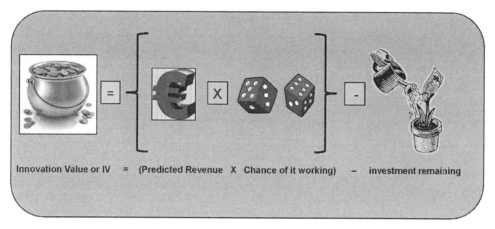

Innovation Value or IV   =   (Predicted Revenue   X   Chance of it working)   –   investment remaining

where:

- **IV** = Innovation Value
- **Predicted revenues** = This is the potential prize of the initiative in terms of revenues generated or costs avoided
- **Chance of it working** = This is the average of the commercial and technical likelihood of success, expressed as a decimal
- **Investment remaining** = This is the cost (less grants or tax credits) of bringing the innovation to the market which cannot be recovered from the initiative once it starts to spin out cash. That is the sunk costs associated with the initial project spend.

Record possible future project revenues associated with the successful completion of the identified innovation initiatives and their calculated IVs.

In business, financial metrics remain the most commonly-used measurement of the success or otherwise of an innovation or NPD programme. Measurements such as return on investment (ROI), return on capital employed (ROCE) or internal rate of return (IRR) are used to value the outputs of innovation activities. These are usually calculated net of grant funding (front end) or tax credits (back end) supports.

However, I think that it is too restrictive to attempt to measure and value the effectiveness of a business's innovation activities based on financial metrics alone. Therefore, I have given a great deal of attention over the years to the non-financial means by which you can put a value on your endeavours.

In some ways, it is easier to value an innovation output once it has been developed. It is infinitely more difficult to put a value on what has not been completed yet or is in the pipeline. The way we have done this is by measuring the number of potentially successful ideas we collect and making an educated or informed estimate on the sales potential of a typical initiative.

Table 8: Recording projects in your portfolio

Project Description	Project Horizon	Project Type	Estimated Project Cost €	Potential revenues €	IV
Active Projects					
1	3	R&D	-	-	-
2	3	R&D	-	-	-
3	3	NPD	-	-	-
4	2	NPD	-	-	-
5	1 (JDI)	NPD	-	-	-
6	3	BD	-	-	-
7	2	BD	-	-	-
8	1 (JDI)	BT	-	-	-
9	1 (JDI)	BT	-	-	-
10	2	BT	-	-	-
Enabling Projects					
1	1 (JDI)	NPD	-	-	-
2	1 (JDI)	NPD	-	-	-

where: **BD** = Business Development; **BT** = Business Transformation; **IV** = Innovation Value; **JDI** = Just Do It; **NPD** = New Product Development; **R&D** = Research and Development.

Comment: Valuing the innovation pipeline

Company – TellLab, Carlow

Employees – 20

Sector – Laboratory services

Comment by – Mark Bowkett, Managing director

As the managing director and CEO of our business, I report to a board and must justify all our investment in innovation to our shareholders. I am therefore duty bound to scrutinise all our proposed innovation projects in order to make an educated call on the risk : reward profile of our programme. I am a scientist by qualification but very much a businessman at heart. I find the business element to be the strongest part of my own decision-making process when marginal decisions have to be made.

Our company procedures require each innovation project to be analysed by conducting a 'commercial impact measurement' (potential revenue and profitability) on the proposal. This is sometimes as straightforward as a simple return on investment calculation (ROI) but other metrics are needed for mid-term or longer-term innovations where ROIs are difficult to measure or where other factors need to be taken into account.

As R&D is a new venture in our company, we have committed to a five-year plan with the backing of the shareholders. We are ensuring that this commitment is spread over a range of short, medium and long-term projects so that a sustainable and robust R&D pipeline can be developed. These projects have different risk : benefit profiles and we are careful to categorise the projects as follows. Three levels of innovation projects are recognised and different approaches and measurements are needed for each category in order to analyse performance:

- Short-term / medium risk with immediate returns – 3 months to 12 months
- Medium-term / medium risk projects with foreseeable future returns
- Long-term / high risk projects that may not yield quick returns but will be building blocks for future commercial activities.

The source of funding and the amount of resources plus time we are prepared to commit to varies with the level of risk : benefit involved and is evaluated in each case. It is important to recognise that the company's activities in the medium and high risk projects would be very limited without external support.

The table below summarises our approach to funding relative to the risk : benefit of a project:

Risk : reward profile	Funding source	Partner
Short-term / Low to medium	Self-funding from retained profits	No
High risk / high rewards	National funding	No / Yes (Enterprise Ireland)
Very high risk / very high rewards	European Funding	Yes (Horizon 2020 or Life+)

8: THE ROLE OF STANDARDS

Innovation is not the product of logical thought, although the result is tied to logical structure. (Albert Einstein)

I am a big supporter of accrediting any system developed, as it keeps everyone on their toes. There are several initiatives happening that may lead to the development of an integrated innovation management system either at European (CEN) or international (ISO) level. In the meantime, the current state of play regarding standards in innovation management is as follows.

National standards

Innovation management has been on the agenda of the National Standards Authority of Ireland (NSAI) for some time. In 2009, this led to the publication of a guidance note to business in Ireland on the steps required in setting up a formal SWiFT (Specification Written in Fast-Track).

The Innovation Management Standards Committee (IMSC) was established in 2009, as a voluntary national mirror committee of NSAI (TC 45) and is responsible for co-ordinating and promoting the development of standards in the area of innovation management in Irish business and support organisations. Its primary role is to monitor, and where possible actively participate in, the development of international (ISO TC 279) and European (CEN, TC 389) standards in innovation management. It has a key role in representing Ireland's national interest in this area through 'experts' from the group attending meetings on the topic and reporting back to the committee for guidance. A supporting role for the IMSC is to promote the benefits of enterprise adopting the innovation standardisation process.

The group's vision is:

To assist Ireland become a world leader (exemplar) in innovation management through the development, implementation and maintenance of standardisation in the sector thereby enhancing our reputation and creating and sustaining economic wealth for all our citizens.

European standards

A brief history of the process in Europe is that the CEN Technical Committee (TC) was created by CEN in November 2008 in line with the CEN internal objective "to encourage more standards to support a culture of innovation in Europe". The scope of this committee is "the standardisation of tools that allow companies and organisations to improve their innovation management". In summary:

- Secretariat and Chairmanship: AENOR (Spain)
- Regular participants: Austria, Belgium, Cyprus, Denmark, Finland, France, Germany, Ireland, Netherlands, Norway, Portugal, Spain, Sweden, UK
- A number of plenary meetings have been held to date.

Six working groups have been established to look at the various elements of innovation management as follows (CEN 16555):

- Part 1: Innovation Management System
- Part 2: Strategic Intelligence Management
- Part 3: Innovation Thinking
- Part 4: Intellectual Property Management
- Part 5: Collaboration Management
- Part 6: Creativity Management
- Part 7: Innovation Management Assessment

All of the above parts will be published as CEN Technical Specifications (TSs). Part 1 was published in 2013; parts 2 to 6 in December 2014; and part 7 is expected in 2015.

International standards

Established in 2013, the remit of ISO 279 is similar to the CEN initiative. The stated scope of the ISO team is the standardisation of terminology, tools, methods and interactions between relevant parties to enable innovation. The ISO standards technical committee currently has 21 participating countries and 11 observing countries. It too has established a number of working groups to look at different aspects of the subject:

- ISO / TC 279 / WG 1 Innovation management systems
- ISO / TC 279 / WG 2 Terminology, terms and definitions
- ISO / TC 279 / WG 3 Tools and methods.

ISO TC 279 is currently setting up a new working group to look at innovation management assessment.

However, in the absence (for the time being) of any internationally recognised innovation standard (CEN or ISO), accreditation to the accepted international system ISO 9001:2008 can be achieved.

ISO 9001:2008 is the current version of the ISO 9001 series and provides a set of standardised requirements for a quality management system (QMS). With over 1,000,000 organisations currently certified to the standard globally, it is the fore leader not only in quality management systems, but also within all management systems.

The QMS standard is adaptable to any size of company – large or small; multinational or local; public or private – as it offers the same scalable benefits for each organisation.

The overall objective of the standard is to help companies to meet statutory and regulatory requirements relating to the product, while achieving excellence within their customer service and delivery. It also can help to grow market share, driving costs down and managing risk more effectively. The standard can be used throughout an organisation to improve performance within a particular site, plant or department.

The ISO 9001 standard provides a framework of globally-recognised principles of quality management, including: customer focus, leadership, involvement of people, process approach, system approach to management, continual improvement, factual approach to decision-making and mutually beneficial supplier relationships. These are also known as the eight key principles of quality management.

Some companies are implementing a QMS to ISO 9001:2008 for the innovation process. ISO 9001:2008 promotes the adoption of a process approach when developing, implementing and improving the effectiveness of any QMS and so it fits with the process approach that can be developed in the review gate innovation process. Certification to the standard will ensure that a systematic approach is taken to innovation management throughout an organisation, thereby improving its performance.

There is a real opportunity for the commercial sector on the island of Ireland to use innovation standards (especially CEN and ISO standards) to leverage tangible market benefit for our products and services on the world stage.

> **There are several initiatives happening that may lead to the development of an integrated innovation management system either at European (CEN) or international (ISO) level.**

Figure 32: Certification of an IMS to ISO 9001

THE INTERNATIONAL CERTIFICATION NETWORK

IQNet and NSAI hereby certify that the organisation

for the following range of activities

The management of the innovation process within the Innovation Centre.

has implemented and maintains a

Management System

which fulfills the requirements of the following standard

I.S. EN ISO 9001:2008

Registration Number:	IE-19.5172
Registration Date:	24 June 2010
Last Amended on:	24 June 2010
Remains valid until:	23 June 2013

NSAI

Signed: Signed:

René Wasmer   Maurice Buckley
President of IQNet   CEO NSAI

Issued on 01 July 2010

The validity of this certificate is maintained through on-going surveillance inspections.

National Standards Authority of Ireland, 1 Swift Square, Northwood, Santry, Dublin 9, Ireland

IQNet Partners*:
AENOR *Spain* AFNOR Certification *France* AIB-Vinçotte International *Belgium* ANCE *Mexico* APCER *Portugal* CISQ *Italy*
CQC *China* CQM *China* CQS *Czech Republic* Cro Cert *Croatia* DQS Holding GmbH *Germany* DS *Denmark* ELOT *Greece*
FCAV *Brazil* FONDONORMA *Venezuela* HKQAA *Hong Kong China* ICONTEC *Colombia* IMNC *Mexico* Inspecta Certification *Finland*
IRAM *Argentina* JQA *Japan* KFQ *Korea* MSZT *Hungary* Nemko AS *Norway* NSAI *Ireland* PCBC *Poland*
Quality Austria *Austria* RR *Russia* SII *Israel* SIQ *Slovenia* SIRIM QAS International *Malaysia* SQS *Switzerland* SRAC *Romania*
TEST St Petersburg *Russia* TSE *Turkey* YUQS *Serbia*
IQNet is represented in the USA by: AFNOR Certification, CISQ, DQS Holding GmbH and NSAI Inc.
* The list of IQNet partners is valid at the time of issue of this certificate. Updated information is available under www.iqnet-certification.com

IQNET (NSAI branded) NL A4 (2)

Comment: Standards in IMS

Organisation – NSAI (in particular the Innovation Standards Committee), Dublin

Employees – Over 140 staff in the NSAI but the Innovation Management Standards Committee (IMSC) is made up of just one Standards Officer on a part-time basis with the remainder of the committee made up of volunteer experts from Industry

Sector – Standardisation

Comment by – Maurice Buckley, CEO, NSAI

Since the mid-2000s, Irish government policy has been firmly focused on building intellectual capital and promoting investment in the innovative product, process, and service development central to a successful 'knowledge economy'. Thus whether as part of a project consortium within the FP6 / FP7 / Horizon 2020 programmes or individual company projects, there have been many Irish organisations, large and small, embarking on a new journey and engaging in significant levels of innovation and product development. There is a lot of public and private investment involved. We have to get this process right.

Can guidance be provided on managing the innovation process itself? Yes. Can this be couched to address the particular 'innovation' environment pertaining in Ireland? Yes. How best to capture and disseminate these good practice guidelines with a national *imprimatur*? A standard! Therefore in 2008, at the request of Enterprise Ireland, NSAI quickly brought together a high-powered group of experts in this field from industry and academia who volunteered their time to sit on a group tasked to develop said standard. In the spirit of all this innovation and conscious that a normal standard development time of two years would be prohibitive, NSAI developed a new standards product itself for this project called a SWiFT (Specification Written in Fast Track). SWiFT 1: *Guide to Good Practice in Innovation and Product Development Processes* was completed within four months and published in March 2009!

The *ad hoc* expert group developed since into the Innovation Management Standards Committee (IMSC) with this documented vision:

To assist Ireland become a world leader (exemplar) in innovation management through the development, implementation and maintenance of standardisation in the sector thereby enhancing our reputation and creating and sustaining economic wealth for all our citizens.

Therefore the group's primary aim is to help private commercial activities and state-assisted bodies that are tasked with training, business development and the promotion of innovation. The primary role of the IMSC is to assist these stakeholders and in so doing to nurture and facilitate the effective adoption of innovation standards in the economy. These stakeholders will benefit from a structured ecosystem for the management of innovation with tools, practice aids, expertise, and incentives for wide and successful adoption in both private and public sectors. There is considerable proof that innovation across the business and including R&D enables disproportionate growth and job creation. A disciplined approach to innovation therefore is critical: where this is in place, growth and jobs follow. As in other areas of management, the use of standards enables a more disciplined and more effective approach to the innovation process.

Standards enable improved thinking, greater focus, and use of proven practices. NSAI is uniquely in a position to influence and interpret the international standardisation efforts currently under way (by CEN and ISO) that will give valuable guidance on best practice in innovation management.

My genuine wish is that the group, with the assistance of the NSAI, will promote the use of international standards and support Irish-based enterprise with a range of appropriate guides to aid their application in the adoption and exploitation of innovation in their business. I fully support the IMSC's endeavours and urge you all to become involved in it.

9: EXTERNAL FUNDING SUPPORT

It's tough when markets change and your people within the company don't. (*Harvard Business Review*)

To fully exploit the opportunities available, companies need a clear understanding of all the R&D funding supports that are available to them. However, often it can be time-consuming to become familiar with the different supports available. To date, there has been a lack of clear information on the wide range of assistance available and a perception that it is difficult to access support (this problem is particularly the case for start-ups).

Regardless of the source of funding, there is a concept in innovation management which states that the time when it is most difficult to attract external funding and where most innovations fail is after the period of the initial set-up when the seed capital and national grant funding has been spent and before the true commercial of the invention is visible.

Figure 33: The funding 'valley of death' for successful outputs of innovation

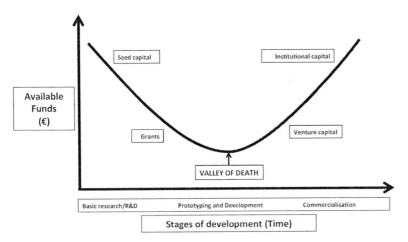

If a business plans to invest significantly in targeted applied innovation projects over the coming years, then it may be eligible for external grant funding or a Corporation Tax credit for this investment. Essentially, the funding support can be indigenous (Republic of Ireland or Northern Ireland) or from Europe. The following section attempts to summarise the current situation.

General sources of direct grant funding: Republic of Ireland

Enterprise Ireland

Enterprise Ireland (EI) is the state agency in the Republic of Ireland responsible for supporting the development of manufacturing and internationally traded services companies. It provides funding and supports for companies or individuals in a number of ways. It is advisable to contact EI directly before embarking on a grant application for assistance from them. There is funding support available from EI to established SMEs in the area of innovation management, entrepreneurship and business transformation, including:

- In-company and collaborative R&D support
- Supports to enhance and develop a management team
- Productivity and business process improvement supports.

Enterprise Ireland is responsible for supporting High Potential Start-Up (HPSU) companies, which are start-up businesses with the potential to develop an innovative product or service for sale on international markets and to create at least 10 jobs and €1m in sales within three years.

There is also support available from EI for larger companies in the area of innovation management, entrepreneurship and business transformation, including:

- Market research and internationalisation supports
- Supports to enhance and develop a management team
- Productivity and business process improvement supports
- Company expansion packages.

In line with its commitment to generating economic value from publicly-funded research, the Government has provided funding to establish industry-led Technology Centres (TCs). These centres are collaborative entities established and led by industry. They are resourced by highly-qualified researchers who are market-focussed and concentrate on strategic R&D for the benefit of industry. This is a joint initiative between Enterprise Ireland and IDA Ireland, allowing Irish companies and multinationals to work together in these centres.

There are currently approximately 15 industry-led research centres in the programme, ranging from bio-refining / bio-energy to applied nanotechnology and pharmaceutical manufacturing to cloud computing. EI also has established and funded a number of technology gateways throughout the country, mainly in the Institutes of Technology (ITs).

Finally, a recent innovation in EI is the Innovation 4 Growth Programme. This programme is targeted at ambitious companies seeking to drive growth through innovation. It supports companies through an end-to-end innovation learning and practice journey to deliver on one or more innovations.

Science Foundation Ireland

Science Foundation Ireland (SFI) is the national foundation for investment in scientific and engineering research. SFI invests in academic researchers and research teams who are most likely to generate new knowledge, leading edge technologies and competitive enterprises in the fields of science, technology, engineering and maths.

SFI makes grants based upon the merit review of distinguished scientists or engineers. SFI advances co-operative efforts among education, government, and industry that support its fields of emphasis and promotes Ireland's ensuing achievements around the world. SFI also provides grants for researchers from around the world who wish to relocate to Ireland and those already based in Ireland, for outstanding investigators, for conferences and symposia, and for collaboration with industry. Proposals are evaluated in open competitions *via* a combination of international peer review and strategic fit with SFI's mission.

SFI also funds a number of major strategic research centres in Ireland that involve several aspects of a national system of innovation agenda. There are currently over a dozen research centres, ranging from biopharma and medical devices to photonics and marine renewable energy research. In addition, SFI currently also funds a number of centres for science, engineering and technology (CSET) research in universities, as well as strategic research clusters in the life science, information/ communication and emergent technology (IC&ET) fields.

Sustainable Energy Authority of Ireland

SEAI has a stated aim of transforming Ireland into a society based on sustainable energy structures, technologies and practices, with a vision of making the country a recognised global leader in sustainable energy.

If you are or intend to be involved in the development or piloting of new technologies in the energy management space, SEAI may be of direct assistance or point you in the correct direction of who can offer you financial support.

IDA Ireland

IDA is the Irish government agency responsible for overseas investment. Since 1994, IDA Ireland has focused exclusively on the promotion and development of high-quality foreign direct investment (FDI) in Ireland, in the manufacturing and international services sectors.

IDA accepts that continuous innovation plays a central role in Ireland's future as a knowledge-based economy. Its campaign promoting Ireland as an innovative, dynamic and educated place to invest is a cornerstone of its promotions. Recognising this, the Irish Government put in place a national Strategy for Science, Technology and Innovation (SSTI) over a decade ago. From this significant funding and support are available to fuel innovation across industry, research and education.

In addition, in the Republic of Ireland, Enterprise Ireland and IDA jointly finance and administer the technology centres (of which there are currently 17), which are industry-led research bodies professionally run by expert researchers but comprising of fee-paying commercial members who largely set the research agenda. The TCs are attempting to break the traditional suspicion between competitors by carrying out industry-specific innovations.

InterTradeIreland

InterTradeIreland is an organisation that has been given responsibility by both North and South Governments to boost North / South economic co-operation to the mutual benefit of both jurisdictions by encouraging better use of our collective resources. It helps to expedite trade and business growth across the island and to create an environment where it is easier to do business.

InterTradeIreland supports SMEs across the island to promote identify and develop North / South trade and innovation opportunities. In addition, InterTradeIreland funds targeted training for qualifying business MDs and CEOs, which allows SME leaders to transform their company's prospects in just nine months by learning and applying the most effective methods of generating, marketing and launching new products and services (the key elements of IMS).

Environmental Protection Agency

The EPA has a statutory role in co-ordinating environmental research in Ireland. Its Science, Technology, Research and Innovation for the Environment (STRIVE) research programme has been driven by national environmental protection regulations and similar European Directives. The STRIVE programme consists of three key pillars: Water, Climate and Sustainability. A series of research calls are announced periodically over the course of the STRIVE programme.

Teagasc

Teagasc is the agriculture and food development authority in Ireland. Its mission is to support science-based innovation in the agri-food sector that will underpin profitability, competitiveness and sustainability in the business.

Teagasc, through its research, has a long tradition of adding significant value to Irish and international organisations through access to expertise, resources, infrastructure and / or intellectual property. Access to technical and consultancy services and commercialisation opportunities at Teagasc is managed through its Technology Transfer Office (TTO).

County Enterprise Boards and Local Enterprise Offices

Ireland's County & City Enterprise Boards (CEBs) were set up in 1993 to stimulate economic development and to cultivate an ethos of local entrepreneurship and local commerce. The CEBs were established primarily to develop indigenous potential and to stimulate economic activity at local level primarily through the provision of financial and technical support for the development of small enterprises. Since 2014, however, the CEBs have been discontinued and replaced with an local authority-based support model for small business, now called Local Enterprise Offices (LEOs).

The LEOs' brief is to support the development of micro-enterprises at local level. The LEOs can support individuals, firms and community groups provided that the proposed projects have the capacity to achieve commercial viability. They can provide both financial and non-financial assistance to a project promoter. The forms of financial assistance that are available, subject to certain restrictions, include capital, employment and feasibility study grants. Non-financial assistance is also available.

Údarás na Gaeltachta

The overall objective of Údarás na Gaeltachta is to ensure that Irish remains the main communal language of the Gaeltacht and is passed on to future generations. The authority endeavours to achieve that objective by funding and fostering a wide range of enterprise development and job creation initiatives and by supporting strategic language, cultural and community-based activities.

A range of financial incentives is available in the form of grant assistance to assist varied business needs. In addition, if a business is located in the Gaeltacht, and wishes to explore a business opportunity or obstacle then Údarás na Gaeltachta can allocate an Innovation Voucher worth €5,000. This voucher can be exchanged for advice and expertise from participating knowledge providers (universities, Institutes of Technology or publicly-funded research bodies, etc).

General sources of direct grant funding: Northern Ireland

The *Innovation Strategy for Northern Ireland 2013-2020* recognises that innovation is one of the primary drivers of economic growth, underpinning the growth of the best performing regional and national economies across the world. Innovation enables firms to stay ahead of competitors, and with global economic conditions remaining challenging, the focus on innovation is now more important than ever.

The *Innovation Strategy* identified the following key market sectors where Northern Ireland has both the capability and the potential to compete on a global basis, and therefore it will prioritise funding and support for research and innovation in both its education and company base in these priority areas: advanced engineering (transport); advanced materials; agri-food; life and health sciences; ICT; telecommunications; and sustainable energy.

Small Business Research Initiative

The Small Business Research Initiative (SBRI) is a well-established process to connect public sector challenges with innovative ideas from industry, supporting companies to generate economic growth and enabling improvement in achieving government objectives.

SBRI provides innovative solutions to challenges faced by the public sector, leading to better public services and improved efficiency and effectiveness. It generates new business opportunities for companies, provides small and medium-sized enterprises (SMEs) with a route to market for their ideas and bridges the seed funding gap experienced by many early stage companies. It supports economic growth and enables the development of innovative products and services through the public procurement of research and development (R&D).

Invest Northern Ireland

Invest NI's role is to grow the local economy. This is achieved by helping new and existing businesses to compete internationally, and by attracting new investment to Northern Ireland. It is part of the Department of Enterprise, Trade and Investment and provides strong government support for business.

InterTradeIreland

See above under Republic of Ireland.

General sources of direct grant funding: Europe

EU Horizon 2020

Seen as a means to drive economic growth and create jobs, Horizon 2020 has the political backing of Europe's leaders and the Members of the European Parliament. They agreed that research is an investment in the future and so put it at the heart of the EU's blueprint for smart, sustainable and inclusive growth and jobs, boosting national economies on the international market, and strengthening the basis for sustainable prosperity and employment.

EUREKA

EUREKA is an intergovernmental network launched in 1985, to support market-oriented R&D and innovation projects by industry, research centres and universities across all technological sectors. EUREKA individual projects are market-oriented international R&D projects supported and funded through direct EU funding. Projects in any kind of technology can receive support on the basis of the quality of a business plan. The ultimate goal of this programme is to ensure that the innovation reaches the market place.

The Eurostars Programme

This is a European Joint Programme dedicated to R&D-performing SMEs, and co-funded by the European Community and EUREKA member countries. Eurostars aims to stimulate these SMEs to lead international collaborative research and innovation projects by easing access to support and funding. It is fine-tuned to focus on the needs of SMEs, and specifically targets the development of new products, processes and services and the access to transnational and international markets.

Eurostars projects are collaborative, meaning they must involve at least two participants (legal entities) from two different Eurostars participating countries. In addition, the main participant must be a research-performing SME from one of these countries.

A Eurostars project must be market-driven: it must have a maximum duration of three years, and within two years of project completion, the product of the research should be ready for launch onto the market. The exception to this rule applies to biomedical or medical projects, where clinical trials must be started within two years of project completion.

Marie Curie

This is an EU funding initiative that allows eligible research scientists or engineers at all stages of their careers, irrespective of nationality, to participate. The programme supports researchers working across all disciplines, from near-term process innovations to 'blue-sky' science. The Marie Curie programme also supports industrial doctorates and promotes the combining of academic research study with practical commercial investigations in companies to enhance employability and career development.

LIFE programme

LIFE is the EU's financial instrument supporting environmental, nature conservation and climate action projects throughout the EU. Since 1992, LIFE has co-financed some 4,171 projects, contributing approximately €3.4 billion to the protection of the environment and climate.

LIFE+

The new LIFE Regulation was published in the Official Journal of the European Union in 2013. It establishes the Environment and Climate Action sub-programmes of the LIFE Programme for the next funding period, 2014 to 2020. The budget for the period is set at €3.4 billion.

To fully exploit the opportunities available, companies need a clear understanding of all the R&D funding supports available to them.

For more information on the funding support available specifically to your business, contact the organisations listed in **Chapter 9** directly.

Comment: Funding and support for your IMS

Organisation – Industry Research and Development Group (IRDG)

Employees – 3

Sector – Support, education, networking, funding and the provision of a representative voice of industry in innovation management

Comment by – Denis Hayes, CEO

The IRDG is a non-profit, business-led Innovation Network of member companies, working together to drive excellence in innovation within Ireland's industry to create growth, jobs and prosperity. Established by industry in 1992, IRDG is an independent body serving the needs of members on all matters relating to research, development and innovation (RD&I). Membership is relatively evenly divided between Irish-owned and overseas-owned companies, which range in size from start-ups to the largest companies in Ireland. IRDG represents all sectors of industry, including electronics, software and telecommunications (ICT), financial services, food, software, engineering, healthcare and life sciences, plastics and utilities. It also includes a significant number of research-performing third-level institutes. IRDG is principally funded by members' annual subscriptions.

IRDG operates under five pillars of activity:

- Representation
- Funding and support
- Innovation networking
- Collaboration
- Learning.

Under the funding and support pillar, IRDG helps member companies maximise funding for their innovation and R&D through state and EU grants and R&D tax credits. It is becoming ever more apparent that continued investment in RD&I is not just desirable but highly necessary to continue to expand Ireland's businesses and economy. Given the array of R&D grant funding available, it can be difficult for many business owners to take the time to collate the data from multiple sources and to identify the most appropriate funds to target. The processes can appear complicated and demanding. However, with the right support and advice, companies can avail of and maximise the significant supports that exist.

There are a number of funding mechanisms that support in-house R&D, including grants and R&D tax credits. Increasingly, both state and EU incentives are geared towards collaboration between industry and the higher education institutes. IRDG has experience over 20 years of supporting member companies to identify and secure the most appropriate grant funding for their initiatives. Member companies have received between €175m and €200m in grant funding since IRDG was set up in 1992. Member companies also maximise their tax incentives through IRDG's involvement and support on R&D tax credits. IRDG works on a number of levels to help its members to access funding incentives:

- It uses all of this knowledge and experience to collate on its website a summary of funding opportunities for companies of all sizes

- It organises seminars and workshops throughout the country to keep members abreast of the many, varied and evolving funding incentives that are available. Expert speakers present at these events and are drawn from state agencies, large accounting firms, consultants and member companies

- It also provides one-on-one support to companies ranging from basic information and advice, through search for collaborative partners to full grant proposal preparation

- It promotes sharing of experience and knowledge between member companies

- IRDG has an on-going dialogue with state agencies, thus providing constructive feedback from member companies on how the various schemes are performing and suggestions for new and improved supports.

Tax credits for your innovation spend: Republic of Ireland

Republic of Ireland

The R&D tax credit scheme has been highly effective in increasing business investment in R&D in Ireland since its introduction in 2004. Over the years, a number of improvements to the scheme have occurred so that Ireland remains competitive as a location of choice for R&D activities, as well as a means to stimulate an increase in such activities by indigenous companies.

In addition to the front end acquisition of direct grant project funding, all businesses operating an innovation programme in Ireland need to ensure that they achieve the maximum permitted tax credits for all expenditure made in these and related areas. Changes in the 2010 to 2013 budgets increased the level of tax credits to 25% for all qualifying spend.

All claims in this regard are levied against comparable expenditure, if relevant, in 2003. This is known as the base year and the tax credit due is calculated by reference to it (from 2015 the base year will be removed all together).

What activities are covered by the process?

All qualifying projects or activities must demonstrate the following to be included:

- **Systematic, investigative or experimental activities:** Must be planned, logical and methodical. Detailed records must be maintained. For example 'R&D projects' with technical goals, progress reports and results, managed and directed by suitably qualified experts
- **In a field of science or technology:** Natural sciences / Engineering / Medical sciences / Agricultural sciences
- R&D must fall under one of the following categories:
  - *Basic research:* New experimental work undertaken without specific practical application in mind
  - *Applied research:* Possible commercial applications for basic research findings
  - *Experimental development:* New or improved products / devices for commercial application.

All spend must seek to achieve scientific or technological advancement and involve the resolution of scientific or technological uncertainty.

What records do you need to keep and what oversight will there be?

An R&D tax credit claim is made via the company's Corporation Tax return and must be made within 12 months of the accounting year end of a company.

If a company has not made a claim to date, it is very important that great care is taken to ensuring the calculation for the 2003 base year is correct, accurate and verifiable. This figure cannot be reduced or changed, apart from in rare circumstances, and therefore this figure will form the foundation of any future claims in this regard. Some bodies are lobbying to have the base year removed.

Strict regulations are in place to ensure potential claimants can demonstrate adequately a systematic and organised approach to qualifying projects and activities. It is vital that going forward all new, and potentially qualifying R&D projects follow a pre-defined and 'standard' model of organisation and structuring. There is no standard recording template used by Revenue. However, there are a few things to take note of:

- Project reports do not have to be submitted to Revenue unless expressly requested
- The 23-point Revenue Query letter should be used as a reference of what is required
- Presentation on the project should be made by the engineer or a scientist / technologist and not an accountant
- An external expert may be used by Revenue; you will be told about this beforehand
- Full interaction between the R&D department or the nominated third party and finance throughout the process is critical.

Scientific record-keeping

Companies are required to maintain and record source documentation to satisfy the science test.

Accounting record-keeping

Companies are required to maintain and record source documentation to satisfy the accounting test.

Sooner or later, your business will have a Revenue audit if it has made R&D tax credit claims. In any event, one can expect a Revenue audit at least once every three years and possible more often.

Preparation for a Revenue audit

As the saying goes, fail to plan then plan to fail! By routinely recording all of the technical and financial information referred to in the 23-point Revenue Query letter and keeping the technical files up-to-date, you will be well prepared for the audit. Remember, it will happen sooner or later so be prepared. It is worth repeating that the content of the audit is both financial and technical, so keep both sets of records and involve staff from both disciplines. If all or part of the work conducted is outsourced to a third party, then have them provide your financial department with the information necessary to support your tax claim.

Presentation at a Revenue audit

It is vital that Revenue is fully aware of what your business does, its position in the market place, how it carries out innovation or R&D and how your business benefits or will benefit from it. A formal presentation by the business leader on this is a good idea. This should be followed by details of the claim by the technical people involved (addressing the science test criteria) and the financial staff to back up this with data (addressing the accounting test criteria). If an external expert is employed by Revenue to assess the claim, he / she will lead the audit and is likely to focus mainly on the technical validity of the claim. So have all your technical people and external advisors on standby!

A word of caution!

Anecdotally, I have heard some things in my research for this book that give cause for concern in the area of tax credits. It seems that the sometimes over-rigorous approach taken by Revenue in conducting these audits may deter businesses from applying for tax credits into the future. This would be counter-productive, as the aim of the process is to encourage the commercial sector to innovate. It is contended that the current adversarial process is at risk of becoming anti-innovative.

It is important to say that all the businesses I spoke to are fully supportive of tax auditing *per se* and all are most reputable and fully tax-compliant in all areas. In addition, many of the companies have always been most diligent about the preparation and maintenance of project defence files (both technical and financial) – the maintenance of good records is something they take great pride in. However, there were a few exceptions, particularly in the trading SME sector. Without exception, all of those consulted agreed that it pays to have the technical and / or the commercial people liaise closely with the finance department in your business in order to ensure that all angles of the process are covered.

Where tax audits were conducted, the companies were first contacted by Revenue to inform them that an R&D tax audit was scheduled and that a reputable technical expert (usually an academic from a university) in the area would technically assess the projects that had originally qualified for funding under Revenue's rules. Many of the businesses interviewed were not overly concerned at this stage and were confident of the scientific rigour and calibre of their R&D staff. In a number of cases, this all changed on conduction of the audit. Over a series of days, the staff involved were required to comprehensively defend the science and the accounting aspects of their claim. There were many discussions and document exchanges during the audit based on technical and financial queries from the Revenue auditor. While many of the exchanges brought clarity to enable effective decisions by the auditor on the topics in hand, as the audit progressed it appeared that there was an underlying objective for the auditor to reduce the amount of the tax credit awarded.

There are a number of nuances to the tax credit obtained that called into question the future of the continued participation of these businesses for the 25% of all qualifying spend. Many of the companies I spoke to were supported via the expertise of the Industry Research and Development Group (IRDG). In addition, a number of private consultants are active in this sector who, for a fee, advise on applications and even manage the reporting and returns for some. All felt that it was expensive enough to make a claim for the 25% credit on spend. This additional element of the process left many asking whether it was really worth it.

My wish is that the audit team and its associated processes are consistent with the original objectives of the tax credit system of encouraging businesses to innovate (with its associated risks of failure) to drive commercial competitiveness for the benefit of the business and also the national economy. All policymakers recognise that R&D is a critical element of our innovation and business development agenda going forward. Let's make the auditing process thorough – but not excessively rigorous on participants.

Tax credits for your innovation spend: Northern Ireland

A scheme very similar to that run by Revenue in the Republic of Ireland also exists in Northern Ireland run by HMRC. To claim R&D relief, simply include qualifying minimum expenditure in your company tax return and you will benefit on your usual Corporation Tax payment date for the period. The R&D Relief Scheme is a key element of national government policy and is now the largest single source of government support for business R&D.

Comment: How we maintain records in order to obtain funding and claim tax relief for our R&D activities

Company – Arran Chemicals , Roscommon

Sector – Pharmaceutical

Comment by – Anthony Owens, CEO

Arran Chemical Company is an independent fine chemicals business located near Athlone, Co Roscommon. We manufacture products for pharmaceutical and health care, flavour / fragrance, personal care, and other specialised chemical and industrial applications. Besides these specific product ranges, we also manufacture other chemical intermediates for existing pharmaceuticals and for new chemical entities, for including monomers and dyes for contact lenses, cosmetic chemicals and photographic solvents.

Over the last 30 years, the company has built up a comprehensive capability covering a wide range of reaction technology, especially focusing on chiral chemistry and innovative organometallic reaction chemistry. Our reaction technology is supported and reinforced by a continuous programme of investment in R&D, leading to significant process innovation and technical improvement.

The maintenance and auditing of comprehensive record keeping is a core element of all we do. It is not hyperbole to say that we could not exist if we did not properly manage our records. This is even more critically important when Arran works with our clients in partnership arrangements to secure external grant support for the programmes we are working on. In addition, we use our meticulous record-keeping process to archive auditable evidence required to gain the maximum tax relief on our R&D efforts.

We use our externally appointed financial auditors to record all the necessary records to satisfy the Revenue's 'financial test'. Our technical / scientific R&D files are maintained by the laboratory scientists themselves (the 'scientific test'). Our auditors summarise this technical information for inclusion in our Corporation Tax returns each year. All the required backup information is kept in our offices.

We find the funding support that we or our partners get through direct grant aid or tax credits for allowable spend invaluable and we are doing more innovative things as a result of this aid. We believe that the maintenance of good records of our endeavours in this area is a small price to pay for this..

Finally, we fully expect to have a Revenue tax audit on our previous year's claim soon. We are ready and well prepared for this event because of our sound record-keeping!

10: RISK MANAGEMENT

If you're not failing every now and again, it's a sign you're not doing anything very innovative. (Woody Allen)

Key risks – and how to manage them

Whatever can go wrong, will go wrong – but do not panic, it can be fixed! **Tables 9 to 14** identify the top six potential risks in setting up and running a formalised IMS, the implications of not managing the risk and suggested mitigation measures to minimise or eliminate the risks.

Table 9: Risk 1: Sufficient senior management support is not forthcoming

ISSUE	IMPLICATION	MITIGATION
All stakeholders in the business do not see the relevance of what is being done.	Non-participation of key staff in the process as they cannot see its relevance.	Be relevant by choosing projects based on market insights! Make sure the innovation process is clearly aligned to the vision and strategy of the business.
Business leaders fail to sustain meaningful levels of innovation, particularly in H2 / H3.	The effort is not maintained due to the entire process being seen as out of touch with the present, not relevant and too futuristic in nature.	Operate a balanced portfolio approach to innovation management.
Failure to sustain the process because key influencers in the business do not see its real value.	No value is placed on the innovation projects in the pipeline.	An Innovation Value (IV) or similar metric should be developed to assist the businesses set a priority for the innovation. Also such a metric helps maintain a balanced approach to your company's efforts.
Failure to bring people in the business along with you on the journey.	There is little knowledge of the process outside a small group of stakeholders.	Do not be shy about communicating successes.
Failure of the process can result by losing the ability to sustain a strong and effective innovation process across your business.	Management is unwilling or unable to fix what is broken or underperforming.	Review the process often and examine the on-going KPIs and CSFs set.
It proves difficult to regain a robust level of innovation throughout the business once it is lost.	Loss of interest in the process, resulting in the loss of central control and a 'free for all' process gaining a foothold in your business.	Keep the process fresh by holding regular seminars for participants and have outside speakers to sense check how your company compares.

Table 10: Risk 2: Inability to take the plunge and invest in innovation

ISSUE	IMPLICATION	MITIGATION
Risk aversion takes hold, thereby curtailing the innovation practices.	Process is confined to lower risk H1 projects.	Do not be afraid to fail: remember less than 1% of what you try will be commercially successful.
Poor communication of what this initiative is about.	Innovation can lose support from the main population and become seen as the remit of a few who are removed from the market place and reality.	Identify the key influencers in your business and involve them in the process.
No ideas are submitted and there is little visible engagement from employees in the system's operation.	Innovation in all three horizons ceases due to a shortage of ideas and possible projects.	Make it known that no ideas or initiatives are too small (do not scare people off submitting ideas).

Table 11: Risk 3: Correct level of funding is not assigned to innovation

ISSUE	IMPLICATION	MITIGATION
Scarce funds are diverted to other more pressing frontline business needs.	Innovation is curtailed due to lack of funds.	Obtain the maximum external grant funding and corporate tax credits for the company's innovation management process.

Table 12: Risk 4: Failure to align the IMS agenda to the market's requirements

ISSUE	IMPLICATION	MITIGATION
System is not sufficiently customer-centric.	Customer or market insights are not taken into account when setting the innovation agenda.	All ideas and projects must be based on firm market insights.
A clear alignment of the innovation strategy as an enabler for the entire business to reach its vision is not documented.	Company vision and innovation strategy are not aligned, resulting in conflict.	The PESTEL and SWOT approaches can be used to regularly review your innovation objectives in order to ensure that the programme being delivered is well-grounded in market insights.
Failure to conduct innovation activities based on market insights.	Process seen as being removed from the customer and the market place.	Focus on solving customer needs – be based on market insights.

Table 13: Risk 5: The innovation approach taken is not ambitious enough

ISSUE	IMPLICATION	MITIGATION
The business is exposed to a disruptive or radical innovation.	Too much focus on continuous improvement or incremental innovation, leading to exposure to the business to a disruptive or breakthrough market initiative putting you out of business.	Keep a diverse range of projects underway throughout the business spanning the horizons and looking at initiatives from incremental innovation to radical developments and from the H2 and H3 areas.
Too concerned with incremental innovation.	Too narrow a focus.	Don't always focus on the here and now; look to the future and be market-aware as opposed to market-led; also operate a balanced portfolio.
Process becomes closed from outside inputs.	Efforts become focused on the here and now and lose sight of the future.	Look for inputs outside the immediate company by operating an open innovation model with as broad an ecosystem as possible.
Wrong people are chosen for the team.	Process becomes too confined and involves just a small number of individuals.	Innovative ideas should come from many sources. Leadership from the top is needed to support the change.
All decisions regarding innovation are centralised.	Innovation is seen as the remit and role of a few people and removed from the business's realities.	Anyone and everyone should be permitted to innovate and participate in the process; a de-centralised or hybrid structure helps to stimulate this.
The process is not accredited to a national or international standard.	Process lacks credibility.	Gain accreditation for the system to an internationally-recognised standard.

Table 14: Risk 6: Try to do too much too soon!

ISSUE	IMPLICATION	MITIGATION
Results do not materialise as expected.	Enthusiasm for the process is lost by key influencers.	Innovation is on-going (it is a journey not a destination); manage expectations!
Try to do too much and impress too many.	Loss of focus with attendant stretching of resources and budgetary overruns.	Don't spread yourself too thin!

These risks should be formally assessed and scored on a monthly or quarterly basis in order to ensure that the innovation agenda is being correctly addressed and that it remains aligned to the overall business objective.

A formal innovation risk management process

I have found it very useful to operate a formal risk management system to keep us on track. There are many risk management models that can be used. The following is one such approach:

Risk is defined as an event that threatens the achievement of organisational objectives.

The theoretical model may be described as follows:

- **Inherent risk:** Inherent risk is the pre-control impact and likelihood of risks materialising. It is described as follows:

Likelihood X Impact = Inherent Risk

- **Control:** Control is any activity that prevents the likelihood of a risk event happening or reduces the adverse effects of a risk on the achievement of organisational objectives

- **Residual risk:** Residual risk is the exposure remaining to the business after the effects on the identified inherent risks have been considered. It is described as follows:

Inherent Risk – Controls = Residual risk (exposure)

Ideally, an on-going monitoring (auditing programme) should be introduced in the business following the formal risk assessment exercise. On a bi-monthly or quarterly basis, the degree to which the management function is complying with the controls identified in the formal risk assessment process should be measured.

Critical controls model

This system works on the assumption that keeping on top of controls for identified operational risks will limit or eliminate risk exposure.

Table 15: Critical operational controls

Critical Factor	Impact (I)	Likelihood (L)	Inherent risk (I x L)	Controls (C)	Residual risk (I x L) x (C)
Lack of senior management support for the process					
No value is placed on the innovation pipeline					
Too much focus on incremental or H1 innovation, exposing the business to being made obsolete by a disruptive or breakthrough innovation					
Too much focus on disruptive or radical innovation, leading to taking the eye off core business activities					
Market insights are not considered					
The system fails to demonstrate value by not setting realistic KPIs					

Scoring:

Inherent risk Score	Residual Risk Score (exposure)
(I x L)	(I x L) x (C)
1 Low inherent risk	1 Low level of exposure
3 Medium inherent risk	3 Medium level of exposure
5 High inherent risk	5 High level of exposure

Comment: Risk management in an IMS

Company – Innovation Centre, IBM Technology Campus, Dublin

Employees - > 1,000

Sector – ICT applications in modern life

Comment by – Noel Crawford, Manager

IBM is one of the world's leading providers of advanced information technology, products, services and business consulting expertise. We are dedicated to helping our clients innovate and succeed through the end-to-end transformation of their business models and the application of innovative technology and business solutions. Innovation is what we are all about. In fact innovation management is one core value that is embedded in our culture. We openly say that we are about 'innovation that matters – for our company and for the world'.

At IBM Ireland, we began to sense the new possibilities for innovation several years ago, as we worked with our clients to help them become more agile, more responsive and more adaptive in their business practices. This idea of innovation management is nothing new to us. We have always been an enterprise-focused innovation company. To get those discoveries to profitable and beneficial life in business, government, education and Irish society at large has always been our primary aim. We call this 'innovation that matters'.

IBM Innovation Centres, in Dublin and elsewhere, offer research and development expertise to enable Irish start-ups and small to mid-sized companies create leading-edge software and hardware applications. The Dublin Centre also provides a platform for accelerating innovation in business by providing insights into new and emerging technologies. The Centre is co-located with the IBM Executive Briefing Centre, The Dublin Software Lab, The IBM Watson European Competency Centre and the IBM Design Studio on our Dublin 15 site, thereby giving our partners access to a deeply-skilled resource pool. In addition, we also engage openly with business partners, academic leaders and IT professionals and entrepreneurs in the field so as to create a fully inclusive innovation assessment network.

The IBM Innovation Centre offers the resources start-up partners need, including free use of IBM software on the cloud for test and development purposes, technical expertise, marketing and sales facilities and skills, and IBM cloud and platform services. Partners may collaborate with IBM and other business partners through our worldwide centres. One of our main activities is that we encourage ideas and proposals from our academic partners, our team of motivated and very innovative employees and from our clients (partners) who, after all, are those closest to the market insights. We do not discourage ideas from any source.

As we partner with our clients to develop and implement a robust risk management system for their businesses, it is logical that we operate a strict risk regime ourselves in our approach to innovation management. Risk: reward determinations are therefore a critical element of us managing the Innovation Centre's activities in Dublin.

We adopt a comprehensive list of principles which we adhere to in evaluating the efficacy of ideas proposed by our partners and we subsequently operate a strict approach to the project management of any initiative which progresses. This also includes a formal and robust approach to project risk management.

11: REVIEW

An idea that is not dangerous is unworthy of being called
an idea at all. (Oscar Wilde)

On a regular basis, it is good to step back and seriously assess the successes of the innovation process; otherwise, the project can lose support from critical stakeholders. It is useful to monitor the performance of the system against agreed critical success factors (CSFs), which may include some or all of the following:

- Has buy in from all groups been secured (especially from senior management)?
- Have the business's vision and strategy changed since the last review?
- Are our innovation objectives still fully in line with our overall business vision and strategy?
- How much are we spending on innovation – and is this enough?
- Is the structure for the management in the business still appropriate for our business or do we need to amend it?
- How are the innovation team performing and does it need to be changed or refreshed?
- Has the value of the innovation initiatives in the pipeline being calculated?
- Has the business case for the current and future spend on innovation being made?
- Does the business operate a balanced portfolio approach to its current innovation management process?
- Has the maximum funding support in terms of direct grants or tax credits for our innovation efforts been secured?
- Have the benefits of the process been shown to operational units and customers?
- Do participants agree that we have created and operate a simple and user-friendly innovation management system?

- Is the ideation (coming up with ideas) system in place working well?

- How many ideas were processed and from where did they come (internal *vs* external)? Were they market pull or technology push ideas?

- Is the system for capturing and prioritising of ideas submitted working well?

- Is a formal project management process for assessing ideas that are prioritised in place and, if so, how is it working?

- Do we operate a review gate approach to project management in accordance with the 'fail quickly and fail cheaply' concept?

- Have we successfully brought to the market and commercialised any of the initiatives?

- Do we operate communications programme on our initiatives and, if so, is it effective?

- Do we operate an open approach to innovation and, if so, what participation do we get from stakeholders outside the company?

- Are all of our activities based on market insights?

- Do we communicate the function of innovation well to internal and external stakeholders?

- Did we focus our initial efforts in areas so as to achieve 'early wins'? What comes next?

- Is there transparency in the process and appropriate feedback to participants?

- Do we have a formal risk management process in place for our innovation management process?

- Do we actively manage expectations so as to avoid loss in confidence of the process?

Comment: Reviewing our IMS

Company – Fastank Engineering Ltd, Antrim, Northern Ireland

Sector – Specialist tank manufacture

Comment by – Seamus Connolly, CEO

We are a very successful, fast erection tank manufacturing business based in Northern Ireland but with a global sales spread. Innovation (both near- and long-term focus) has been the secret to our success. It is part of what we are, part of our DNA. It is fun to meet and overcome the challenges that lead to innovation. This is where we get our corporate buzz!

We do not currently operate a formal IMS in our business, therefore no official review of this process in our organisation occurs. However, the topic of NPD and our R&D reaction to current or future market changes are very critical to our business. As part of our annual planning process, we look at the market and assess changes or developments as they may affect our trading position in the coming year. Our NPD response to these changes in order to protect and / or extend our market share is regularly assessed and openly discussed. The marketing and sales people at the customer coalface are at the forefront of such discussions. Although we very much adopt an 'open' approach to our innovation activities, collaborating with the likes of Queen's University Belfast (QUB), much of our work is done in-house by our resident design engineer.

In addition to the annual planning process, there is an informal review or assessment done on each project we embark on. This is conducted on a case-by-case basis but, as is the situation with most SMEs we do business with, we do not have the resources in-house to do a formal review or an interim appraisal on how we are progressing with our entire innovation agenda. On the contrary, we review how individual projects are progressing continuously. Our company's motto in our innovation activities is to 'fail quickly and fail cheaply'.

The starting point for all our innovations is the market place. Here, we aim to establish, by heavy reliance on market insights, the requirements (product or service development) of our existing customers or of new customers (those we do not have yet). Once we establish this, it is progressed on to whether the requirement of the market can be done or not (technical evaluation). Only then will we commit time and other resources to a project.

We are small enough to respond quickly if a particular development project is going off track or indeed needs to be discontinued because it has not worked or is no longer required in the market place. However, we do expect the whole innovation process to become more formal as we grow. My hope is that we do not lose the flexibility to react quickly and decisively to market insights as we do now.

12: INNOVATION IN PRACTICE

> The five essential entrepreneurial skills for success are
> concentration, discrimination, organisation, innovation and
> communication. (Michael Faraday)

At the time of writing, Ireland's goal of becoming the best place in the world to do business by 2016 may well be on track to become a reality. According to *Forbes'* ranking of the Best Countries for Business published in 2014, Ireland scores top across the board. New Zealand ranks second overall; third is Hong Kong; fourth and fifth are the Scandinavian countries of Denmark and Sweden respectively.

As we all know only too well, the Irish economy was devastated by the Great Recession (mid-2000s to 2014 – we hope it is over!). Yet despite these economic troubles, Ireland still remains attractive to overseas investors. The world's biggest companies have established here over the past decade. Our capital city, Dublin, now serves as the European headquarters for a number of US tech firms including Google, EBay, PayPal, Amazon, LinkedIn, and Facebook.

Given the recent woes our country has experienced it is truly wonderful to see us top of the bunch when it comes to doing business but there is no room for complacency. We need to remain attractive to indigenous business and FDI alike. Innovation management is a key factor in this.

The timing for this publication could not be better! Ireland is becoming an exemplar in this area. Let's keep it that way.

Innovation closer to home

In my conduct of the research for this book, I have met with many very progressive and highly innovative companies throughout the island of Ireland. These have ranged from SMEs employing just a handful of people to larger indigenous companies turning over billions of euro on the world stage and to the much-admired FDI multinationals that are active in this country in the high-tech ICT and pharma sectors. It is interesting to note that the smaller the enterprise, the less likely the business is to recognise its activity as innovation. I have been told "we do not innovate; we develop our business model in response to market change". In many of these situations, I have asked "Is your business different now than it was 10 years ago?". "Totally", is the prompt response I usually get. I rest my case!

Republic of Ireland

There is little doubt that the multinationals operating in Ireland are world class innovative companies and that their efforts in this area are very much driven by market insights. The likes of Google, Facebook, IBM, Intel, EBay, PayPal, HP, Pfizer, etc. are indeed exemplars in this area. In many cases, these giants either have a dedicated R&D function or an NPD programme in place (or in some cases both) in Ireland that responds to current or imminent market changes. They operate a process of near-term and longer-term innovation management to agreed percentages in a balanced portfolio manner.

However, I was pleasantly surprised to see that innovation management is increasingly becoming a core activity in many of our large indigenous companies (including our service industry). Similarly, there seems to be a mood for more customer service and business model innovation in the public sector.

I am afraid it is more hit or miss in the SME sector where, understandably, credit management, cash flow and making sufficient margins to stay in business are the most critical things on the owners' minds. Having said this, I did speak to some very innovative and market-aware SMEs.

I encountered basically three types of business. First, there were those who did not see the value in innovation at all and very much had the attitude that, if they did not toil to keep the business they had, they would be out of business in a very short timeframe, so it was the here and now that was their concern. Second, there were businesses that concentrated their efforts on the very short-term. They did not see the value in expending resources towards the future (in many cases, they did not even consider the immediate future). Third, there were highly innovative, market-centric businesses that are fully aware of market changes and constantly looking for new angles where they can progress their business. They are often high-potential start-up businesses and are aware of the limited size of the home market, often focusing on exports to generate future sales. Many of these businesses are closely supported by Enterprise Ireland.

On the government incentives for businesses to carry innovation (R&D, NPD or BD) activities, there is some considerable anecdotal evidence that the policy is working. This is particularly the case for the larger FDI or large indigenous enterprises. Unsurprisingly, larger businesses tend to operate a partial or totally formal innovation management system, while SMEs tend to be much less formal and often informal. They typically prefer 'coffee dock' innovation and can respond rapidly to change, kill projects / initiatives or start new ones. The innovation agenda in these SMEs is often driven by one or a few individuals and succession planning can be an issue.

Northern Ireland

Northern Ireland is well populated with innovative high-potential start-up businesses. Many of the people I spoke to considered that innovation in response to market changes was the most important thing on their agenda. In some respects, I was reminded of the dynamism I encountered in SMEs based in Germany. There was a 'can do' attitude that I felt was refreshing.

In many cases, there was not an over-reliance on grants or on tax rebate incentives to get started. In fact, one business went so far as to indicate that the government incentives, while welcome, were often bureaucratic to obtain and resulted in the loss of independence on their innovation activities, preferring instead to spend much of their hard-earned trading profits on innovating for their future. In this particular case, the funds committed to innovation activities amounted to greater than 14% of the company's revenues – well above the average spend I encountered in my research. This is typical of SMEs' approach to entrepreneurship as they are often impatient to get moving and could not be bother waiting for grant assistance. In fact, many of the start-ups I spoke to in NI preferred the tax credit scheme as a government incentive to their innovation endeavours. It was interesting also that the experiences of certain SMEs in NI with respect to tax credits were markedly more positive than their counterparts in the

South. This may be due to the fact that the scheme is in operation for a longer period in the South and many of the recipients therefore have had a Revenue audit (see my comments earlier).

I met larger indigenous corporate and multinationals too in NI. As a rule, these organisations tend to be more structurally innovative and more organised in how they obtain their market insights. This is primarily because of the greater availability of resources in such organisations but, as is the case in the South, the trade-off is that these businesses tend to be less responsive and more bureaucratic than their SME cousins.

Summary observation

I was disappointed, saddened and frustrated, in equal measures, to find that there is little or no inclusion of formal innovation management instruction in either the North's or South's basic education systems. I am convinced that there is urgent need for this type of training for our students if we are not to yield ground to other countries that are currently rolling out or are contemplating such a system. There are many third level courses in innovation and entrepreneurship but, irrespective of their chosen discipline, all students should study innovation and entrepreneurship in a practical and applied but straightforward manner. The training should be generic to suit both SMEs and larger companies.

International best practice

Israel

> In Israel, a land lacking in natural resources, we learned to appreciate our greatest national advantage: our minds. Through creativity and Innovation, we transformed barren deserts into flourishing fields and pioneered new frontiers in science and technology. (Shimon Peres)

One of my first impressions of the Israeli people is that they are all very proud of what they have achieved in their country in a very short time. They are all well-versed on how this has been achieved – they all agree that innovation is at the core of their success. Innovation is a way of life: there is general acceptance that, through product and service innovation, plus the leveraging of standards to promote same, the country is among the most successful in the world at generating successful start-ups.

Developing a culture of innovation in the commercial sector is key to the country's success. Every person consulted on innovation management in the commercial sector in Israel (even taxi drivers!) was aware of this.

Like so many other countries, the Israelis' main focus includes energy management, water supply / conservation and clean technologies.

The main driver to commercial innovation in Israel is security of supply. Ironically, the geopolitical implication of Israel being situated in a region that is largely hostile to its very existence has been instrumental in promoting an innovative approach to the generation of ideas, technology testing and market development. For example, the development of rapid testing methodologies for the microbiological cleaning of water supplies (including cryptosporidium) was accelerated by the national need to protect water supplies from sabotage.

Do not be fooled by the ancient majesty of Israel. It is a modern and vibrant country that has embraced the art of innovation management to become a truly world leader in the area. Tel Aviv is a thriving innovation hub in a modern and vibrant city. It is definitely one of the commercial capital cities of the region

Figure 34: Contrasts: Jerusalem and Tel Aviv

Germany

Germany has a deserved reputation of being a reliable, high-tech and very innovative nation. Many of the world's most recognised inventions were discovered in Germany. From the humble denim jeans to the X-ray medical diagnostic tool and from the bicycle to the helicopter, they all came from German innovators. German companies have known for some time that price alone is not everything in getting customers to purchase your product or service. Germany's renowned history for engineering excellence means that consumers buying German goods typically are looking for that little bit extra: be it cutting-edge technology or that special, perfectly-designed something that simply cannot be found elsewhere.

German innovation often is based on very deep technical expertise, which allows even small German firms to become market leaders in very narrow segments. Getting inventions off the drawing board and into the market – true innovation – is one of Germany's key strengths. In addition, the management of intellectual property (IP) and protection of brand reputation are critical to its innovation management agenda. The provision of state aid to innovators is also a critical reason for Germany's success.

As a country, Germany is very keen to practice the process of open innovation. The government actively encourages its researchers to network internationally. It has set up centres for research and innovation in many other countries. The government also supports R&D with solid financial backing.

The most important challenge for Germany is to remain in its current strong position in the global economy. There is an acceptance that this will be achieved primarily by innovation.

China / Hong Kong

The Chinese government's innovation strategy is all about the country urgently looking for ways of designing and making its own products rather than manufacturing someone else's as it does currently. The Chinese government therefore has committed a large portion of its ever-growing wealth to R&D and innovation. There is little doubt now that high-tech innovation is shifting from the Western world (particularly USA and Germany) to China. However, when it comes to turning discoveries into useful new products, the hallmark of innovation, Chinese businesses still seem to struggle. China is responding to this challenge by developing a high-tech domestic workforce and in so doing becoming a significant driving force of the global innovation in industry. In the biotechnology sector, this is especially true: the Chinese government recently has embarked on a Five Year Plan, which aspires to lead the country's economy towards a high-tech era with a special focus on biotechnology.

Hong Kong, now a region of China, is also making a major play for creating a dynamic economy based on innovation. A new regional government-backed bureau has been established to conduct a thorough audit in order to ascertain the current status and needs of the industry in the area. It is hoped that this will lead to a clearer roadmap from which resources and policies can be developed. A substantive plan would encourage more investment, not just in infrastructure but also in human resources.

Norway

'Innovation Norway' is the Norwegian government's official trade representative abroad and its most important instrument for innovation and development of national enterprises and industry.

It is now well-accepted that Norway possesses a number of key features when it comes to promoting innovation in business:

- It builds competitive Norwegian enterprises in both domestic and international markets
- The country's enterprises are well promoted on the world stage
- The centres of innovation excellence are well spread out across the country and rural areas are very much included
- It focuses on transforming ideas into successful business cases and bringing same to the market
- It promotes interaction between enterprises and R&D institutions (open innovation).

The goal of Innovation Norway is to promote nationwide industrial development with a clear focus on both the individual business and Norway's national economy. To achieve this, Norwegian enterprises have access to a broad business support system as well as financial means. The government offers direct funding support but also provides non-financial, advisory services, promotional services and network services.

Finland

It is indisputable that Finland has an innovation-driven economy. In fact, innovation management policy in Finland has been the key pillar of successive governments for many years now. Development of Finland's innovation system is co-ordinated by the Research and Innovation Council, led directly by the Prime Minister.

Finland's innovation system has been ranked among the best in the world. Competence-based competitive advantages are chiefly employed in support of the national economy and securing well-being, by boosting business's competitiveness and growth.

In addition to direct financing, it is accepted at the highest levels that there are many factors that influence businesses' willingness to innovate. These include legislation, access to international markets, and the functioning of the EU's internal market.

In the public sector, a number of measures are in place to encourage engagement in innovation activity. Since the public sector does not produce goods for markets, the goals of its innovation activities differ from those of the private sector: among other tasks, the public sector seeks to provide citizens with more useful public services as well as improved productivity in public sector activities.

United States of America

There is little doubt that R&D is the backbone of the success that the competitive, knowledge-driven USA has enjoyed for many years now. There is an acceptance at federal level in the USA that R&D investment helps to develop new products and services that drive growth, create jobs, and improve national welfare. For this reason, the US government traditionally has spent more than any other nation as a percentage of GDP on directly funded R&D.

In addition, the federal government also promotes innovation and R&D through its tax policy. An incentive known as the Research and Experimentation (R&E) tax credit encourages private sector R&D by allowing corporations to take unlimited deductions for qualified research spending.

Another innovative policy gaining traction in the US is the 'patent box', which taxes firms that develop patents at a lower rate on income related to domestically-developed patents.

Privately-funded R&D in the United States is also well-developed. The largest business sectors for US R&D are computer and electronic products, pharmaceuticals, software and computer services, R&D services, automobiles, aerospace and defence.

United Kingdom (UK)

Like in many other countries, innovation management in the UK is promoted at senior government level. The Department for Business, Innovation & Skills (BIS) is the relevant department. It invests in skills and education to promote trade, boost innovation and help people to start and grow a business.

Within BIS, the Technology Strategy Board is the UK's innovation agency. Its role is to stimulate innovation, working with business and other partners, in order to accelerate economic growth. The Technology Strategy Board's vision is for the UK to be a global leader in innovation and a magnet for technology-intensive companies, where new technology is applied rapidly and effectively to create wealth (sounds familiar!).

However, despite having innovation as one of the key strategic pillars of government's policy, recently Britain has been shown to spend less as a proportion of national income on R&D than much of the rest of the EU.

New Zealand

It is accepted in New Zealand (NZ) that all businesses have ideas and knowledge in their lockers that just needs to be identified and released. In order to benefit from these ideas, it is important to ensure that the best ideas are harnessed and translated into new products and services to increase organisational growth and performance. This is key to NZ's government policy on innovation management. Like in many other countries, there is a government department specialising in entrepreneurship and innovation. The Ministry of Business, Innovation and Employment's role is to encourage, promote, assist, fund and support science innovation within the country. Many groups are involved in the Ideas Acceleration programme that helps business to reduce project costs, increase efficiencies and develop new revenue streams – the ideal outputs of innovation!

New Zealand has made many innovative market launches. The jetboat, the ski plane, bungee-jumping, the Zorb ball, terra-sailing and the referee's whistle are examples of New Zealand inventions. New Zealand continues to innovate in a range of fields, from its traditional export industries of agriculture and dairy, to newer growth areas of technology and award-winning wine. All this innovation is promoted, encouraged and rewarded by a robust national government policy – for example, the Innovation Council is a government-led initiative designed to accelerate the commercialisation of innovation by NZ firms.

Figure 35: Some well-known New Zealand inventions

Bungee jumping

Terra-sailing

The Jetboat

The Zorb Ball

The Referee's Whistle

Denmark

Denmark frequently tops many of the world rankings for innovation management and entrepreneurship, within areas like intellectual assets and open, excellent and effective research systems. It is often, however, argued that Denmark's national support for innovation and entrepreneurship is a formula that only functions for smaller countries. Nonetheless, Denmark is an exemplar in the sector and operates a model that other jurisdictions could do well to mimic.

There is a wealth of federal support for innovation in Denmark. If an individual wants to be entrepreneurial, then the state will cover his / her health insurance as well as providing adequate social support to provide the fundamentals of existence, food and shelter as that person takes the 'risk'!

The government assists a business to scout for the technology it needs, promotes its research projects or finds the right R&D partners for them. There is also support available for evaluating the market potential for innovative technology, connecting companies to international networks or reviewing their business plan.

Because Denmark is a small economy, it has to be open to foreign research communities, technology networks and venture capital. International collaboration is considered as essential to ensure national companies, researchers and institutions continually improve their competitiveness. Denmark is currently leading the way in areas such as cleantech, ICT, and life sciences.

Comment: Innovating in a large multinational organisation

Company – Dalkia / Veolia, Dublin

Comment by – Mark Coyne, Innovation director

Developing a systematic innovation competence is difficult in any organisation. It involves leadership, top level commitment, development of a process that is entirely new, winning the hearts and minds of all employees, as well as finding enthusiastic people who can drive, shape and guide the organisation through the process. Every single one of these actions is challenging in themselves, but when you add all that into the context of a multi-national organisation, the challenges compound and multiply and make the overall task of developing the innovation competence much more complex. This piece explores this concept and suggests ways in which these challenges can be overcome.

The nature of a multi-national organisation can be described in two ways: it has multiple layers and a central function. Layers either can be by geographical country / area, by product, by service, or by regional zone. These layers are usually overlapping, so for example service line A is a profit centre that reports into country CEO B, which reports into a zone director C, etc. In terms of developing innovation at a country level skill, the individuals who are charged with responsibility for running these layers are all stakeholders, and it requires skilful and careful management on behalf of the innovation manager. Then there is a central or corporate function, which is usually a pivotal stakeholder, again requiring careful management.

Some key factors that need to be considered in launching a local innovation initiative are:

- Is there strong local CEO / MD support, or is the drive for the initiative coming externally? A local strong sponsor is essential for success, whereas an external supporter is desirable but not essential

- Has the organisation (either locally or corporately) set innovation as a priority, but just does not know how to do it? Often there is vagueness in the support for innovation, but a poor understanding of how to achieve it. This lack of clarity is fine – innovation capability is a difficult concept, but the priority setting will generate goodwill capital as and when support at various 'layer' levels is required

- Is there a command and control culture in the central organisation? In a multi-layered organisation, such a culture is commonplace and is largely driven by a desire for standardisation and / or a need for control (financial or operational). This type of culture is very challenging in which to develop an innovation competence and requires a very strong local culture to counter it.

Another challenge that a potential innovation manager may face is an organisational reluctance to embrace external innovation. Some multi-national organisations favour not just idea generation that comes internally, but also internal idea development, prototyping and deployment. This can limit options and create non-optimal development paths. Some strategies to ensure a successful innovation development programme in a multi-national organisation are:

- Identify innovation best practice: Don't re-invent any wheels – this is a sure recipe to get to a dead-end initiative. Identify where good innovation process practice exists and find out where the gaps are. Copy what's good, and develop solutions to fill those gaps. If there really is no best practice, then you have free rein to create something totally new

- Stakeholder engagement and communication is key: Identify and develop relationships with key stakeholders, especially outside the local jurisdiction

- Start simple and small: Do not overcomplicate the innovation process and keep it readily understandable to those who might be distant from its implementation

- Think pilot: Your local innovation initiative, if successful locally, can be spread as best practice to other parts of the organisation. This is a real win / win for your non-local stakeholders

- Develop an ideas filter that matches the corporate strategy: Idea filtering and what to focus on is key to success – you cannot do everything all at once. Designing a filter that provides this focus is a key way to ensure an aligned outcome both locally and with the central corporate function

- Demonstrate successes and identify quick wins: This helps in selling the concept, and most importantly selling the benefits.

13: YOUR NEXT STEPS

I believe in being an innovator. (Walt Disney)

Where to get help

There has never been a better time for your business to start a formal innovation management process. There are many government departments that are only too willing and able to assist you or point you in the right direction. Foremost in the innovation management process development in the Republic of Ireland is Enterprise Ireland (for companies exporting or expanding outside Ireland) and IDA Ireland (for businesses outside Ireland setting up here), while in NI it is Invest Northern Ireland. InterTradeIreland is a useful port of call if your business has a North / South co-operation dimension. Below are some of the organisations that you can contact to help you set up a system or expand and develop one.

Table 16: Organisations that can help you get started

ORGANISATION	WEBSITE
Advice / Support	
Commissioner for start-ups (Dublin)	www.ryanacademy.ie
Enterprise Ireland (EI)	www.enterprise-ireland.com
Environmental Protection Agency (EPA)	www.epa.ie
Fáilte Ireland	www.failteireland.com
Forfás	www.forfas.ie
IBEC (Innovation, Science and Technology committee)	www.ibec.ie
Industrial Development Authority (IDA)	www.idaireland.com
Industry Research & Development Group (IRDG)	www.irdg.ie

ORGANISATION	WEBSITE
Invest Northern Ireland (INI)	www.investni.com
Knowledge Transfer Ireland (KTI)	www.knowledgetransferireland.com
Local Enterprise Office (LEOs)	www.localenterprise.ie
National Standards Authority of Ireland (NSAI - IMSC)	www.nsai.ie
Sustainable Energy Authority Ireland (SEAI)	www.seai.ie
Training/Standards	
Industry Research & Development Group (IRDG)	www.irdg.ie
National Standards Authority of Ireland (NSAI - IMSC)	www.nsai.ie
SOLAS	www.solas.ie
Funding Support	
Enterprise Ireland (EI)	www.enterprise-ireland.com
European Union	http://ec.europa.eu/research/innovation-union/index_en.cfm
Industry Research & Development Group (IRDG)	www.irdg.ie
InterTradeIreland	www.intertradeireland.com
Local Enterprise Offices (LEOs)	www.localenterprise.ie
Science Foundation Ireland (SFI)	www.sfi.ie
Collaboration	
IBEC (Innovation, science and technology committee)	www.ibec.ie
Industry Research & Development Group (IRDG)	www.irdg.ie
IOTI (Institutes of Technology representative group)	www.ioti.ie
ISME	www.isme.ie
Knowledge Transfer Ireland (KTI)	www.knowledgetransferireland.com
Science Foundation Ireland (SFI)	www.sfi.ie
Small Firms Association (SFA)	www.sfa.ie
Universities Research group	www.iua.ie

Comment: Innovating in an SME

Company – EpiSensor, Limerick

Employees – 5

Sector – Low energy smart sensor communications devices

Case Study by – Gary Carroll, CEO

EpiSensor is one of the world's leading suppliers of easy to deploy, secure and reliable wireless sensors. It was founded in 2007 and is a result of the coming together of three strands of technology that were rapidly developing at that time. These core technologies are sensor networks, cloud computing and mobile communications.

Our USP is that our platform can dramatically increase energy efficiency, reduce costs and improve sustainability. Data produced by our systems can transform the results achieved using traditional monitoring, control and automation systems. EpiSensor's products are trusted by some of the world's largest and most secure organisations. We have lowered costs and reduced environmental impact for Fortune 500 customers in Europe, Australia, Asia, and North America. Innovation is critical to us as we strive to develop to stay ahead of the opposition in a rapidly changing and developing market.

As an SME, however, our approach to this necessary innovation function is radically different to the well-resourced multinationals. We all hear about the Googles, Facebooks, Microsofts, IBMs etc when innovation in commercial circles is referred to in the media. Well, we in the SME world innovate too! However, the provision of grant support and tax credits that we receive for our activities in this area is a godsend! This helps us to remain active in this critical area and to carry out work that we could not otherwise do due to lack of funds. We are often amused when we hear people in larger organisations (I have to admit, like the author!) speak about formal innovation management. Yes, a formal approach often can be more reflective and thorough but it is often full of red tape and process-heavy also! The flexibility and 'fleetness of foot' that being an SME brings to our innovation agenda is, I believe, a critical element of our success. If something needs to be done or changes made, there is little bureaucracy involved. We just get on with it! In my experience, this is certainly not the case with larger corporates where approvals and counter-approvals can get in the way, resulting in considerable time delays and, as a consequence, missing out on market opportunities.

EpiSensor, as an SME, has the best of both worlds. By design, we have adopted a very open approach to our innovation activities. We engage with the outside world by having a network or contacts approach. We are not too precious about whether a breakthrough happens for us or for our partners (mainly the Clarity Centre for Web Technologies, a partnership between University College Dublin (UCD), Dublin City University (DCU) and the Tyndall National Research Institute, based in Cork). We rely heavily on our external innovation partners to do the fundamental or basic research. We provide inputs of a technical nature based on our market insights. We also provide a route to market for the outputs of the teams' efforts, which after all is what innovation is all about.

Innovation is critical to us as we develop and grow. The entire world is our market place, so keeping up with developments in current technology and emerging innovations is vital. The best way to do this is to be part of a larger network of professionals.

> **There is plenty of help out there. Do not put off asking for it, as your business's future may depend on it!**

Finally

If I were pressed to summarise all about this book and IMS in a single diagram, it would be the following expanded innovation framework (**Figure 36**).

Figure 36: An expanded innovation framework

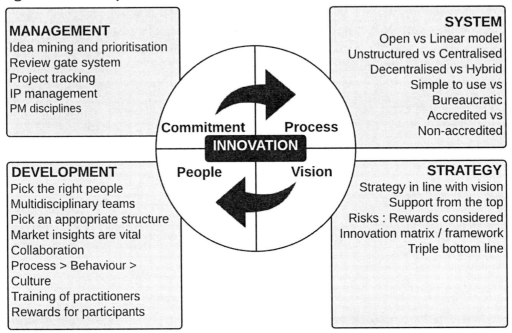

As I hope you have seen, there is no real mystical secret to establishing an IMS in your business. So get started – today! Your business may depend on it!

I am happy that you have reached this stage in the innovation management process – but do not stop here. This is only the first step on your journey. Remember IMS is a journey, not a destination. There is much more you can learn to make you a better innovator. This book just got you started! There are many good training courses and text books available to take your learning further.

> **You are now on the pitch and ready to become a better player.**

Finally, on a positive note, I am confident that Ireland and indeed Europe can continue to innovate and evolve to reach their full potential on the world stage. Do not leave it to others. You are as good an innovator as the next person, so get going!

> **Remember it is the result that counts, not the tools or techniques you use. They are simply enablers to help you on the way we should go or to get you to your destination more efficiently.**

GLOSSARY OF TERMS

If at first, the idea is not absurd, then there is no hope for it.

(Albert Einstein)

I have been involved in other areas of business management where making things complicated with the liberal use of jargon is the name of the game. I prefer to make things easy, so here is a direct and down-to-earth jargon-busting explanation of key innovation terms.

Ambidextrous innovation

This refers to an enterprise innovating in the short-term ('here and now' incremental advances), while at the same time keeping an eye on and committing some resources to the longer term (disruptive or radical innovation). It is said that all businesses should be prepared to be doing something totally different in the future – a total re-invention of self.

Michael L Tushman, a well-respected change management guru, accurately described this approach to innovation management as follows:

> **The 'ambidextrous organisation' describes how companies can pursue breakthrough growth through a two-pronged effort in which they separate their new, exploratory units from their traditional, exploitive ones, while maintaining tight links across units at the senior executive level.**

Analytics

This recent innovation is truly a revolution in the commercial world. Analytics is the collection of consumer or customer data, the conversion of this to a meaningful visual form for the user to make sense of and the communication of this insight to others in the business for commercial advantage.

Firms commonly apply analytics to business data in order to predict future market trends, and to improve business performance by controlling sales price and marketing / promotions, One of the best things about analytics as an innovation management tool is that it is a predictive science in that it may be used to look at future trends as opposed to historical data (looking at your business through the front windscreen as opposed to the rear view mirror!).

Architectural innovation

Architectural innovation is where the form of a product or service offering changes but the basic modules that makes it work remains the same. It is said that the difference between incremental innovation and architectural innovation is that the former is change, while the latter is revolutionary.

Brainstorming

Brainstorming is a group creativity technique or session in which efforts are made to find a conclusion for a specific problem by gathering a list of ideas spontaneously contributed by the members of the group. Brainstorming is usually practiced in an ideation session where emergent and convergent options are recorded.

Business development

Business development is concerned with the conception, development and implementation of any potential growth opportunities for a business. It can have its genesis in diverse areas of a business from R&D, production, marketing, and sales etc.

The conceptualisation of a business opportunity is the first phase (the roadmap) and may come from one or more of these areas of the business but the implementation phase is where the real benefits are recognised and requires all of the sections to work (the business plan).

Business transformation

Business transformation is the total strategic change of a business from what it is doing today into something else that provides for more stability or sustainability in the enterprise. The need for business transformation may be caused by external changes in the market or other external and internal factors such as regulatory change or poor cashflow.

As stated earlier, innovation is a pseudonym for change or transformation and it is through this constant evolution that businesses survive and prosper. It is not a fad that can be taken on in good times and committed to the waiting lounge in hard times. It is something you all need to do all of the time.

Community sourcing

Colleges and universities have used the term 'community source' to refer to a type of community co-ordination mechanism that builds on the practices of open source communities. An important and distinctive characteristic of community source as opposed to plain open source is that, in community source, some organisations or institutions commit resources – human or financial. The plus is that community sourced projects are not purely volunteer efforts as found in other open source communities; the downside is that they may be shaped by the strategic requirements of the institution(s) committing the resource.

Critical success factors (CSFs)

This refers to a limited number (usually between 3 and 8) of conditions, or variables, that have a direct impact on the effectiveness or efficiency, of any business or project. Activities associated with CSF must be performed at the highest possible level of excellence to achieve the intended overall objectives.

Crowdsourcing

Crowdsourcing is the practice of obtaining needed services, ideas or content by soliciting contributions from a large group of people, and especially from an online community, rather than from traditional employees or suppliers. This process is often used to subdivide tedious work or to fund-raise for start-ups.

Chindōgu innovation

From the Japanese word meaning 'weird tool', these are extreme, often comical, innovations with no real purpose or commercial application.

Design thinking

Design thinking is defined as a methodology for practical resolution of problems or issues that looks for an improved future result. It is a form of solution-based or solution-focused thinking that starts with the goal or what is meant to be achieved instead of starting with the problem. It differs from the scientific method, which starts with defining all the parameters of the problem in order to define the solution.

EU 2020 funding

In 2013, political agreement was reached on Horizon 2020, the new EU framework programme for research and innovation funding. Horizon 2020 will play a key role in Europe's response to the economic crisis, as well as strengthening the EU's global position and addressing the challenges that society faces. Horizon 2020 has nearly €80bn of funding available over seven years (2014 to 2020). Its primary objective is to drive economic growth and create jobs. By coupling research and innovation, Horizon 2020 is helping to achieve this with its emphasis on excellent science, industrial leadership and tackling societal challenges. The goal is to ensure Europe produces world-class science, removes barriers to innovation and makes it easier for the public and private sectors to work together in delivering innovation. Horizon 2020 was officially launched by Enterprise Ireland in December 2013 and initial calls for interested parties in five thematic areas were launched soon thereafter.

Funding the 'Valley of Death'

This refers to the likelihood that new innovations of a business will die before a steady and sustainable stream of revenues is established. Unless a firm can effectively manage itself through the Death Valley curve, it will fall victim to a lack of cash flows.

Ideation

Ideation comprises all the processes involved in taking an idea from its initial conception through to realisation of its commercial potential. Ideas may come from a variety of sources, including employees, customers and partners, through channels including planned brainstorming sessions, intranet forums and applications, web-based collaboration tools, surveys and social media websites. Idea management processes typically harvest ideas from such sources and rank them according to a number of criteria. Ideation management processes should encourage full internal and external participation and collaboration (see open innovation).

Ideation session

This is a formal meeting or session where a diverse group of people (usually employees but third parties can participate) discuss a known issue within a business and attempt to solve the problem identified. It is usual for the problem owner to produce an information pack for attendees to examine before the session.

The ideation session is generally facilitated by someone who is not an expert in the chosen area. It is typically organised as follows:

- The redefinition of the issue
- The divergent phase, where the options for a solution are identified; it is vital that all ideas no matter how unrealistic they appear are captured at this stage
- The convergent phase, when technically feasible and commercially viable options to solve the problem are selected from those captured in the initial phase.

An ideation session sometimes is referred to as targeted brainstorming.

Innovation attrition rate

This is simply the rate of project 'die-off' at the defined review gates on the journey through the ideas funnel. Typically only about 1% of ideas reach the market and, of these, less than 10% are a market success.

Innovation ecosystem (also see Open innovation)

This term covers the participants or stakeholders in the innovation or research endeavours of a business. In an open innovation model, it include some or all of the following internal and external participants:

- Employees
- Customers
- Financing and other support institutions
- Entrepreneurs
- Partner companies
- Third level research facilities.

Innovation framework

This is a complicated way of saying that you have evaluated the key elements of an innovation management system (the capture and prioritisation of ideas followed by the delivery of projects and, ultimately, the market exploitation of the resulting outputs). A framework recognises that many inter-dependencies exist between different disciplines involved in the process and therefore different resources are needed to get results.

Inter-dependent elements are recognised as critical in any model as follows:

- Strategy for innovation management must be aligned to the overall company vision
- Commitment from senior management is critical
- The people involved in the process and the organisation of the team
- The process from ideas acquisition to market launch using the review gate model or equivalent.

Innovation funnel

This is a depiction of ideas passing through a process (the funnel) where they are sorted and only a small number of those entering come out the other side to the market place. The funnel is an analogy that depicts a process which captures, filters and prioritises all your ideas. Ideally, a business will run an open innovation model, where it accepts ideas from both external and internal sources. Ideas may originate from a range of sources; but they should enter a single funnel or a group of funnels as raw ideas that are subsequently processed so that only a small portion of them progress to the funnel end – the market. The funnel concept is usually used in conjunction with the review gate process.

Innovation matrix

This is a generic set of innovation categories that you can tailor to your business needs. It is a formal way for you and your staff to identify and document what and where you should be placing your efforts.

Intellectual property

Intellectual property (IP) is a legal concept that refers to creations of the mind for which exclusive rights are recognised. Under intellectual property law in most developed countries, owners are granted certain exclusive rights to a variety of intangible assets – assets that are usually 'virtual' and often cannot be touched or seen. IP can be described as "property of the intellect that can be owned, bought, sold or rented". It comprises the inventive or innovative assets of a business that are developed and benefited from commercially. The main forms of intellectual property are trademarks (usually products or brands), copyrights (written), and patents (usually technology).

Kaizen innovation

The word 'Kaizen' is Japanese for 'improvement'. Its use in business innovation refers to a philosophy that to focus upon continuous improvement of processes in manufacturing, engineering, and business management yields cumulatively the best results. When applied correctly in the workplace, Kaizen can generate some really tangible benefits in small but measurable 'chunks' in all sections of a business. In the best models, Kaizen innovation involves all employees from the CEO to the janitor. The primary aim of Kaizen management is improving standardised activities and processes by the elimination of waste. It is equivalent to Horizon 1.

Key performance indicators (KPIs)

A key performance indicator (KPI) is used to evaluate the success of a particular activity in which a business is engaged. Sometimes, success is defined in terms of making progress toward strategic goals. Choosing the right KPIs relies upon a good understanding of what is important to the organisation. What is important often depends on the department measuring the performance – for example, the KPIs useful to finance may be different from the KPIs chosen by Sales / Marketing. The measurement of KPIs in managing a business is very useful as it often leads to the identification of potential improvements required and is therefore a very useful means by which the success of an IMS can be monitored.

Lean

Lean manufacturing or lean production is a practice that considers the expenditure of resources on anything other than the creation of value for the end customer to be wasteful and thus a target for elimination. Value is defined as any action or process that a customer would be willing to pay for. Essentially, lean is all about preserving value or doing more with less. The principles of lean can also be applied to the development of new products or services.

Linear innovation

The linear model of innovation is an early model of innovation that suggests technical change happens in a linear fashion from invention to innovation to diffusion. One of the failures of this approach has been that it prioritises scientific research as the basis of innovation, and plays down the role of later players in the innovation process. Linear innovation is sometimes referred to as closed innovation and is now seen as outdated.

New product development

New product development (NPD) is the complete process of bringing a new product to market. A product is a set of benefits offered for exchange and can be tangible (that is, something physical you can touch) or intangible (like a service, experience, or belief). There are two parallel paths involved in the NPD process: one involves the idea generation, product design and detail engineering; the other involves market research and marketing analysis. Companies typically see new product development as the first stage in generating and commercialising new products within the overall strategic process but fail to address the second consideration sufficiently.

Non-disclosure agreement

A non-disclosure agreement (NDA), also known as a confidentiality agreement (CA), is a legal contract between at least two parties that outlines confidential material, knowledge, or information that the parties wish to share with one another for certain purposes, but wish to restrict access to or third parties. It is a contract in which the parties agree not to disclose information covered by the agreement. NDAs are commonly signed when companies or individuals are considering doing business and need to understand the processes used in each other's business for the purpose of evaluating the potential business relationship. NDAs can be 'one way', meaning one of the parties shares information with the other, or 'two way' in which both pass on details to each other.

Open innovation

Open innovation is a term promoted by Henry Chesbrough (2003), in his book of the same name. It is a system that assumes that companies can – and should – use external ideas as well as internal ideas. In simple terms, it is "innovating with partners by sharing risk and sharing reward". The main philosophy of open innovation is that there is knowledge outside companies which cannot afford be ignored as to do so is to limit an enterprise's opportunity. In addition, internal inventions not being used in a firm's business should be taken outside the company and sold to raise monies for further innovation. One large sportswear manufacturer has a six-month period during which the company must use the invention, otherwise it will be offered to the market for purchase or licencing. In this way, the business is forced to develop a culture of 'use it or lose it' regarding its inventions.

Open innovation 2.0 is a new concept, currently being developed by the EU. It refers to new innovation ecosystems that include members of society in general as an additional party.

Permanent innovation

This is a phrase introduced by the well-respected expert in the innovation management field, Langdon Morris. It is simply the process of doing innovation continuously as part of the business's routine strategy development. Permanent innovation becomes a habit and a core value of an organisation and thus is very different from random or intermittent innovation in that it happens all of the time and not just when events permit or dictate.

Portfolio management

This is basically keeping an eye on the range of innovation projects or initiatives in which you are involved in order to maintain an appropriate balance between the risk : reward aspect of your business's innovative efforts. Too much emphasis on the high risk but high potential projects can result in the eye being taken off the ball as regards the core business or cash generation. Too much focus on low risk, low return projects can mean that your business is in danger of being fatally damaged by a competitor introducing a disruptor onto the market.

Therefore, any business operating a formalised innovation process should carry out their work in all areas simultaneously depending on their business strategy, business maturity, market share, competition, rate of market change, future plans and, critically, staff / resources available.

Research & development

The primary function of a research and development (R&D) group is to develop new products, improve existing ones, discover and create new knowledge about technological topics of relevance to a company or investigate for the purpose of uncovering and enabling development of valuable new products, processes and services. The vast majority of a company's activities are intended to yield nearly immediate profit or immediate improvements in operations and involve little uncertainty or risk (applied R&D). The R&D discipline in businesses is usually staffed by engineers or scientists.

New product design and development is more often than not a crucial factor in the survival of a company. In an industry that is changing fast, firms must continually revise their design and range of products. This is necessary due to continuous technology change and development, as well as other competitors and the changing preference of customers. Without an R&D programme of its own, a firm must rely on strategic alliances and acquisitions to tap into the innovations of others.

Reverse innovation

In the past, established companies greatly simplified the features of their established products in an attempt to sell these 'no frills' products into significant markets in the developing world (for example, India and China). This approach, however, was found uncompetitive as only the most affluent segments of society in these developing countries could afford the offering, thereby limiting the market potential. Reverse innovation instead develops products locally in developing countries and, once they are successful there, upgrades them for subsequent export to developed markets.

Review gate management

The review or stage gate model for innovation management is a project management technique (first developed by Cooper in 2001) in which an initiative or project is divided into stages (or phases), separated by gates (or hurdles). At each gate, the continuation of the process is decided by a manager or a steering committee (often called gatekeepers). The decision is based on the information available at the time, including the business case, risk analysis, and availability of necessary resources such as money or people. To progress through the gate to the next phase, the gatekeeper must be convinced that there is merit in doing so. It is essentially a systematic approach to innovation management that prevents too much resources being allocated to a particular project until certain criteria are first proven. In simple terms, it is a 'fail quickly and cheaply' system that is designed to develop the maximum numbers of ideas, with the minimum amount of resources and in the shortest timeframe. It does this by assessing technical, marketing and financial criteria as the project progresses from the initial 'quick look' investigation to full market launch.

Systematic inventive thinking

Systematic Inventive Thinking (SIT) is a thinking method developed in Israel in the mid-1990s. SIT deals with the two main areas of creativity and problem-solving. A high rate of ideas per session or event was previously considered an indication of creativity (the numbers game). This approach led to a series of methods for developing creativity-based events on the assumption that a quantitative increase of ideas necessarily brings about a qualitative improvement. Such widely-known methods as brainstorming and lateral thinking (identified with Edward de Bono) can be traced to this approach. However, in recent times, it is considered by many that a large flow of ideas does not necessarily lead to the creation of original ones. A new approach has been proposed that uses organised thinking and structured processes rather than the random generation of ideas. One of the characteristics of SIT is to focus on a small number of realistic ideas. In this approach, originality replaces quantity as a dominant criterion.

The first step in using SIT is to define the problem world. Once defined, the problem-solver knows that all the building blocks for the solution are right there in front of him and that the solution simply requires the re-organisation of the existing objects.

Tax credits in the Republic of Ireland

Since 2004, the R&D tax credit scheme has been a key incentive for attracting and retaining foreign direct investment (FDI) to Ireland, although in addition, it has encouraged domestic R&D activity. The RD tax credit scheme requires a clear identification and documentation of all qualifying R&D activities and expenditure within a business. A 25% R&D tax credit is currently available for companies carrying out qualifying R&D activities in Ireland. Some features of the current system include:

- The credit can be claimed on an incremental basis
- All claims must be submitted within 12 months from the end of the accounting period in which they were incurred
- The R&D tax credit is available to offset against corporation tax (CT). If, however, a business incurs a loss, then the tax credit can be accrued by Revenue or a cash refund issued
- The R&D activities must be carried out by a company in the European Economic Area
- Qualifying expenditure must be less any direct grants.

Tools and techniques used in an IMS

Using tools and techniques may assist your IMS by:

- Assisting with the generation and maintenance of innovation in your organisation, leading to sustained profitable growth
- Helping you look at things in a new way
- Keeping focus (makes ideation relevant to the business)
- Encouraging 'thinking outside the box'
- Getting to the required outcome quicker and often cheaper.

Examples of tools you may encounter in innovation management include:

- Lean
- Six Sigma
- TRIZ
- Mind mapping
- De Bono's 6 Thinking Hats technique
- Lateral thinking
- Brainstorming
- Reverse brainstorming
- Crowdsourcing / Community sourcing
- Balanced score card
- KPIs / CSFs dashboard
- Divergent / Convergent thinking model
- Trend analysis
- Finite element analysis
- Failure modes and effects analysis
- Value engineering.

Triple (or quadruple) helix collaboration

Three bodies are typically represented in the triple helix innovation collaboration model:

- Educational institutions (usually third level)
- The business and commercial sector
- The government and regulatory agencies.

These three collaborators come together in an agreement to carry out some development. In recent years, a fourth party ('society' or 'the customer') has come to be recognised as yet another important element worthy of involvement, thus the quadruple helix collaboration model was invented.

Triple bottom line

The 'triple bottom line' is an acceptance that innovation and sustainability go hand in hand.

> **We should all run our businesses this way because we wish to sustain our enterprise (Profit). We cherish the staff members who make this possible (People). And finally, we conduct our affairs so that we are not harming the very environment that makes this possible (Planet).**

Comment: Getting to grips with the jargon

Company – Richard Keenan and Co. Ltd, Carlow

Employees – >500

Sector – Agri-feed technology

Comment by – John McCurdy, Business Innovation Director

Today ,we are recognised as the most innovative business in our sector and as one of the most progressive privately owned enterprises in the country. Innovation is a way of life for us and we are always aware of our need to continually improve. However, when we started out on our innovation management journey (we did not recognise it as such then!), I must say I was surprised how complicated some people wanted to make this whole business. I found it difficult to get a simple, straight-talking answer to the many questions I had. Our market was changing rapidly and technology was entering our business at a rate of knots. We responded to this by keeping up with the changes occurring at least as much but in many cases better than our main competitors. We also wanted to keep an eye on emerging trends so that we could retain our reputation of being market leaders not followers. That, in a nutshell, was my main objective.

Most of what I read on the subject was, in my opinion, very theoretical and full of unnecessarily complex terms and definitions that were totally impractical to us for what we wanted to achieve. I wanted to get beyond this complexity to develop a simple and easy to relate to objective for our team, which was both commercially justifiable and defendable. We agreed that our primary objectives were to innovate by:

- Making our current service offering to the market better for our customers, thereby protecting and possibly increasing our market share
- Looking at market needs and introduce new products and services to our existing market
- Developing new market outlets for our current and future products.

For a short time, I felt overwhelmed by the subject but quickly gathered myself and committed to break through all this rhetoric and make what we were planning to do much more relevant. I must admit we totally ignored all the jargon documented on the matter and went ahead to do what we wanted to achieve. We were lucky to have a business owner who believed in what we were trying to achieve. This was absolutely critical.

Once we got started, there was no stopping us and all the jargon of reports was duly consigned to the waste heap. There is little doubt that the people who understand your business most are your employees and those who know your products best are your customers. Consulting both of these as often as possible is our secret to success.

We are a very innovative company in our sector. We started on our innovation journey when there was little out there in terms of assistance. We were not put off by the excessive use of jargon by 'specialists' in the discipline. We got on with it in accordance with our customer demands and wishes in addition to firm market insights, and we hope you do too.

EPILOGUE

The world of business is increasingly dominated by the theme of 'customer loyalty' – the magic ingredient that would have your customers wanting to grow the relationship, endorse your products at the expense of your competitors', and often willing to pay a premium for the privilege! Indeed, a higher level of loyalty is demonstrated by those customers seeking to collaborate on new product or process development, while also perhaps investing hard cash in your process. Getting to this point of engagement is approaching the true definition of 'partnership'.

Understanding the concept of 'innovation' is one of the most difficult challenges facing long-established enterprises, many of which came into being before the Internet revolution and often in a market-protected or regulated environment. Many of those organisations and companies clearly see their markets eroding and struggle to successfully migrate their enterprises to the new realities. The much publicised practice of 'disintermediation' has completely disrupted the 'custom and practice' of those enterprises and, together with a staid slow-moving culture, the organisation is doomed to failure before even attempting a fight back!

Innovation, or the conversion of ideas to cash, is a key success factor for all organisations wishing to sustain and endure into the future. The critical question is 'how do we manage innovation?'. Are there key themes that straddle all types of enterprises, whether public or private, small or big, national or global? What about behavioural aspects? Is this a top-down process? Are there cultural barriers to successfully transforming the enterprise and making it relevant, while understanding the once-in-a-generation migration of the world economy that will see Asia, most notably China and India, become centre stage of the global economy for the 21st century?

It was against the background of some of the issues described above that I was appointed CEO of Bord Na Móna in 2008.

Arriving to the peat-based State-owned enterprise just in time for the great recession and financial collapse of 2009-2014, I was struck by the lack of understanding of or resources available to new product or process development. Notwithstanding attempts by some to 'shake the organisation up', whatever innovation was there was well-hidden, a 'rainy day activity' and certainly not applauded. Perhaps associated with that well-trodden public sector malaise of 'making no waves' and staying in one's box, whatever incentives there were to try something new, one had better make no mistakes or one was damned!

Yet the company was facing significant challenges: its overdependence on a reducing peat resource, whose extraction was increasingly being questioned by environmentalists and under threat from more stringent Government and EU regulation, while a poorly-timed move into waste collection was compounded by poor demand in the core Irish market across all the company's products and services.

Following some pretty deep and sometimes painful soul-searching, and the development of a new vision 'A New Contract with Nature', it was time to put some resources and effort into innovation. A Head of Innovation had to be found! To say that Hugh Henry jumped up and kissed me when I asked him to fill the role would be an overstatement! 'Innovation is not a good career move in Bord Na Móna', he told me!

Fortunately for Bord Na Móna (and for me), Hugh accepted the challenge and therein commenced the initial phases of the transformation of the organisation to where it is today: the most profitable and growing renewable energy company in Ireland with a very bright and exciting future!

In **EVERYDAY INNOVATION**, Hugh has provided a comprehensive and practical guide to the 'art' of innovation, and how the process stitches into the fabric of the organisation's vision, mission and values.

While much of the narrative can clearly be identified with what is called the 'open innovation' model, as practised by Kerry Group and others, the book comes alive with practical tips gleaned from Hugh's own experience of leading innovation in a reluctant but co-operative organisation. Practical guides to such subjects as incentives, grants and tax reliefs, as well as pointers to recognised centres of best practice, make this book a valuable companion for any business leader.

I commend Hugh for his time and patience while pulling together all the strands and connecting points that set this book apart and I recommend the publication to those tasked with innovation and change.

Well done on taking the first step on this journey. I hope you enjoyed the book!

Gabriel D'Arcy
January 2015

BIBLIOGRAPHY

Baghai, M., Coley, S. and White, D. (1999). *The Alchemy of Growth: Practical Insights for Building the Enduring Enterprise*, Reading MA: Perseus Books.

Chesbrough, H.W. (2003). *Open Innovation – The New Imperative for Creating and Profiting from Technology*, Boston: Harvard Business School Press.

Cooper, A. (1999). *The Inmates are Running the Asylum: Why High-Tech Products Drive Us Crazy and How to Restore the Sanity*, Indianapolis: Sams.

Cooper, R.G. (2001). *Winning at New Products: Accelerating the Process from Idea to Launch*, New York: Perseus Group.

Duncan, R. (1976). 'The ambidextrous organization: Designing dual structures for innovation' in Killman, R.H., Pondy, L.R. and Sleven, D. (eds.), *The Management of Organization*, New York: North Holland.

Elkington, J. (1997). *Cannibals with Forks: The Triple Bottom Line of 21st Century Business*, London: Capstone.

Goffin, K. and Mitchell, R. (2010). *Innovation Management: Strategy and Implementation using the Pentathlon Framework*, second edition, New York: Palgrave Macmillan.

Kyffin, S. and Gardein, P. (2009). 'Navigating the Innovation Matrix: An Approach to Design-led Innovation', *International Journal of Design*, 3(1), 57-69.

Mc Manus, J. (2012). *Intellectual Property: From Creation to Commercialisation - A Practical Guide for Innovators & Researchers*, Cork: Oak Tree Press.

McKinsey (2010). *Innovation Blowback: Disruptive Management Practices from Asia*, Quarterly Report, New York: McKinsey.

Morris, L. (2000). *Permanent Innovation*, www.permanentinnovation.com.

Tushman, M.L., Smith, W.K. and Binns, A. (2011). 'The Ambidextrous CEO', *Harvard Business Review*, 89(6) (June), 74-80.

Wojcicki, S. (2011). 'The Eight Pillars of Innovation', *Think Quarterly*, Google newsletter.

INDEX

ABOUT THE AUTHOR

Hubert (Hugh) Henry Ph.D, B.Sc. (Hons), CIWEM, C.WEM., C.Sc., C.Env., M.I.Biol.I.

Having worked for a company producing environmental diagnostic kits and lecturing in environmental protection prior to completion of his education, Hugh joined the semi-state company, Bord na Móna, in 1990, to work on a range of development projects focusing on wastewater treatment.

After stints as Consultant, Principal Consultant and later Chief Operations Officer (COO) and acting business leader within Bord na Móna's Environmental Company (now called Anua, a wholly-owned subsidiary of Bord na Móna plc), he was asked to take on the new position of Director of Innovation and R&D in June 2008. His remit in this role is to consolidate innovation management, with the primary objective of developing and embedding a co-ordinated R&D and business innovation approach across the company.

Hugh was awarded a B.Sc. (Hons) in Environmental Science and Technology from the Institute of Technology Sligo in 1986 and, in 1990, a Ph.D. for his research thesis, The movement and attenuation of domestic wastewater in soils and its subsequent pollution of groundwater sources. He also holds a graduate Diploma in Business Finance (with distinction) from the Irish Management Institute (IMI).

He also is:

- A chartered member of C.WEM., C.Sc., C.Env. and a professional member of I.Biol.I.
- An Irish expert representative of the European Working Group TC165 WG 41 on the development of a recognised European standard for small-scale wastewater treatment systems
- Chairman of the Ireland Active (standards in the leisure industry) awards jury
- An external examiner to IT Sligo on the Environmental Science and Technology courses
- A member of the NSAI working group assisting in the production of European (CEN) and international (ISO) standards in innovation management (IMSC)
- A member of the NUIM innovation in the public service group (PSIL)
- A member of the International Society for Professional Innovation Management (ISPIM)
- A member of the IBEC Innovation, Science and Technology committee
- A recognised expert on open innovation management, who has lectured widely on this topic
- A nominated trustee of the Bord na Móna management pension scheme (GESS)
- A board member of IRDG.

Hugh is married to Marie and has two teenage boys, Alex (16) and Luke (14). In his spare time, he is a fanatical salmon and trout fly angler and also enjoys reading, walking the family dogs (Buddy and Roco) and listening to music.

OAK TREE PRESS

Oak Tree Press develops and delivers information, advice and resources for entrepreneurs and managers. It is Ireland's leading business book publisher, with an unrivalled reputation for quality titles across business, management, HR, law, marketing and enterprise topics. NuBooks is its imprint for short, focused ebooks for busy entrepreneurs and managers.

In addition, through its founder and managing director, Brian O'Kane, Oak Tree Press occupies a unique position in start-up and small business support in Ireland through its standard-setting titles, as well as training courses, mentoring and advisory services.

Oak Tree Press is comfortable across a range of communication media – print, web and training, focusing always on the effective communication of business information.

Oak Tree Press

E: info@oaktreepress.com

W: www.SuccessStore.com / www.oaktreepress.com.